Barium Hexaferrite Films Prepared by the Sol-Gel Method for Nonreciprocal Applications

Vom Promotionsausschuss der
Technischen Universität Hamburg
zur Erlangung des akademischen Grades
Doktor-Ingenieurin (Dr.-Ing.)

genehmigte Dissertation

von

FRAUKE KATHINKA HELENE GELLERSEN

aus

WINSEN (LUHE)

2018

First Examiner:	Prof. Dr.-Ing. Arne F. Jacob
Second Examiner:	Prof. Dr.-Ing. Didier Vincent
Chair of Examination Board:	Prof. Dr.-Ing. Gerhard Bauch

Date of Thesis Defense: December 17th, 2018

Bibliografische Information der *Deutschen Nationalbibliothek*
Die Deutsche Nationalbibliothek verzeichnet diese Publikation in der Deutschen Nationalbibliografie; detaillierte bibliografische Daten sind im Internet über <https://dnb.de> abrufbar.

ISBN: 978-3-86387-953-2
Zugl.: Hamburg, Techn. Univ., Diss., 2018
Dissertation, Technische Universität Hamburg

Choriner Str. 85 - 10119 Berlin
verlag@menschundbuch.de – www.menschundbuch.de

Acknowledgment

Working on this dissertation has been a period of intense learning for me, not only in the scientific arena, but also on a personal level. I would like to reflect on the people who have supported and helped me so much throughout this period. First and foremost, I would like to thank Prof. Dr.-Ing. Arne F. Jacob, who gave me the opportunity to work on a very interesting and challenging topic at his institute. I wouldn't have been able to finish this work without his profound belief in my work and my abilities. I also thank Prof. Dr.-Ing. Didier Vincent for the preparation of the second expertise and Prof. Dr.-Ing. Gerhard Bauch for chairing the examination committee. I would also like to extend my deepest gratitude to my former students Jahn Hoffmann, Adrian Rusche, Felix Klichowski, Janine Kira Scholz, Jannes Peschel, Anita Ochsenfarth, Cem Diker, and Fabian Niehaus for their incredible motivation, interest and tireless work. They were the best possible team to successfully work on this topic. I also thank Dr. rer. nat. Siegfried Voigt from the DLR Space Administration for the smooth organization of the research project that served as the foundation of my thesis. Additionally, I expand my thanks to the project partners for the fruitful cooperation and many helpful discussions. Furthermore, I would like to thank the many people who have supported me in performing various measurements and producing all the test structures that were needed within this work. Especially, Carmen Hajunga, who manufactured countless perfectly similar coplanar waveguides for me. Special thanks to my colleagues and friends at the Institute of High Frequency Technology, who created a convivial place to work. Particularly helpful to me during this time were Thomas Jaschke, Sascha Meyne, Christian Rave, Benjamin Rohrdantz, Nora Meyne, and Jennifer Reuschel. Last but not least, I would like to thank my friends and family for always supporting and encouraging me.

Contents

1 Introduction

The aim of this work is to enable small and lightweight nonreciprocal components, namely isolators and circulators. The main application of isolators is to shield sources from reflections due to a mismatched load, whereas circulators are mainly used to separate transmit and receive signals in radar and communication systems [1]. This application is depicted

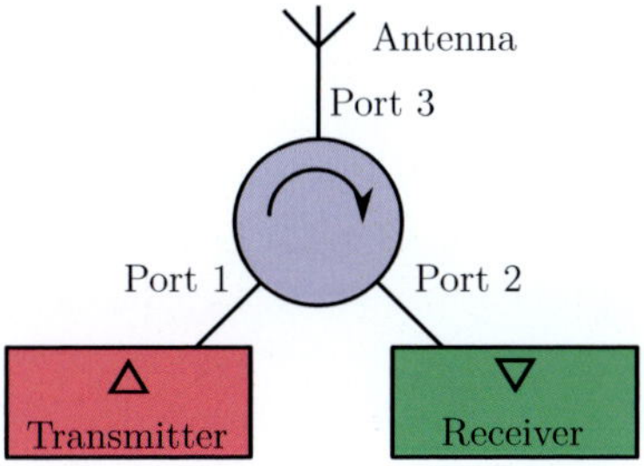

Figure 1.1: A circulator used as a duplexer.

in Figure 1.1, where a transmitter and a receiver share the same antenna and the signals are separated using a circulator. Conventional circulators and isolators using soft magnetic ferrite need a magnetic biasing field, which depends on the operation frequency of the device [2]. For components operating at Ka-band and above, it is very challenging to provide a sufficiently large bias field. Furthermore, the required permanent or electromagnets are bulky and heavy. In contrast to that, self-biased nonreciprocal components rely on the anisotropy field of the material instead of an external biasing field. Therefore, these devices are smaller, lighter and have a higher shock and vibration tolerance. Due to these advantages, it is feasible to integrate these components into low temperature cofired ceramics (LTCC), which offer the opportunity to produce small, low-loss, multilayer microwave components [3].

Aim of This Work

In this work, hard magnetic barium hexaferrite films are produced by means of the so called sol-gel method [4]. Compared to other fabrication processes, this technique is rather simple and low-cost. Furthermore, the method offers the advantage of requiring a comparably low temperature to sinter the final magnetic film. This low sintering temperature enables integration of these films into LTCC. Due to the self-bias capability of the material, which is caused by magnetocrystalline anisotropy, no means of magnetic biasing needs to be integrated along with the component. The anisotropy of the material can be influenced via doping. So, a large frequency range for potential applications can be covered [5]. Especially for future satellite communications at Q-/V-band, self-biased nonreciprocal components, and therefore hard magnetic ferrites capable of integration, are of interest.

Outline

In the following chapter, the advantages of hard magnetic ferrites for nonreciprocal components are discussed with special focus on the material used in this work, namely barium hexaferrite. Different routes that are commonly used to manufacture barium hexaferrite films are outlined and the sol-gel method, which is used in this work, is discussed in detail. Furthermore, the techniques applied for material analysis in this work and the interaction of ferrite materials with microwave fields are briefly presented.

Chapter 3 focuses on the preparation of the barium hexaferrite films. It starts with a brief overview of state-of-the-art barium hexaferrite film manufacturing methods. Afterwards the sol-gel method is introduced. The influence of different process parameters on the produced films is analyzed and suitable parameters are chosen. The effect of doping on the material properties is evaluated and the compatibility of these films with different LTCC tapes is tested.

The last chapter of this work begins with a literature overview over self-biased circulators. Subsequently, a self-biased circulator based on the presented films is implemented and its performance in conjunction with the produced ferrite films is discussed.

2 Basics of Barium Hexaferrite Preparation and Characterization

Ferrites are ferrimagnetic dielectric ceramic materials [6]. In the crystal structure of ferrimagnetic materials the magnetic moments are aligned in two antiparallel sublattices. However, the magnetic moments of the sublattices cancel out only partly and therefore a net magnetization remains.

Conventional nonreciprocal microwave components depend on soft magnetic ferrites in combination with either permanent magnets or electromagnets which supply a static magnetic biasing field. This bias field is necessary to magnetize the ferrite to saturation and to adjust the ferromagnetic resonant frequency (FMRF). The so-called Larmor or precession frequency [7]

$$\omega_0 = \mu_0 \gamma |\vec{H}_0| \tag{2.1}$$

determines the FMRF. Here, γ is the gyromagnetic ratio, μ_0 the vacuum permeability, and $|\vec{H}_0|$ the internal static magnetic field. For high resonant frequencies, a large static bias field is needed. Since microwave components tend to get smaller with higher operating frequencies, it is difficult to provide the required biasing fields in the small area occupied by the ferrite material. Depending on their configuration, conventional circulators are typically commercially available up to 40 GHz for drop-in circulators [1], 35.6 GHz for microstrip circulators [2], and 29.5 GHz for surface-mount circulators [3]. Components built in waveguide technology are available at higher frequencies but are bulky and heavy compared to the other configurations. Furthermore, they are not compatible with planar technology.

[1] JQL Electronic Inc., Drop-In Circulator QR4-700-T-003.
[2] JQL Electronic Inc., Microstrip Circulator JCS34K4T35K6G0.
[3] JQL Electronic Inc., Surface-Mount Circulator JCSM27K5T29K5.

To implement nonreciprocal components at higher frequencies, hard magnetic hexaferrites can be used. Today, circulators and isolators based on hard magnetic ferrites with operating frequencies of up to 40 GHz are commercially available[4]. In comparison to components based on soft magnetic ferrites they offer advantages due to their significantly smaller size and robustness regarding shock and vibration tolerance.

Due to their magneto crystalline anisotropy, hexaferrites have a high anisotropy field which makes them suitable for self-biased applications at even higher frequencies. For example, they can be employed in the Q-/V-band without an externally applied biasing field. In addition a small bias field can be used to shift the FMRF of these materials to even higher frequencies [8]. Even at Ka-band, the permanent magnets used for the bias of the ferrites in conventional components, are large and heavy in comparison to the dimensions of the components. Since no permanent magnets need to be integrated, hard magnetic ferrites simplify the integration of nonreciprocal components. For future satellite communication at frequencies in the Q-/V-band, these materials offer the possibility to design small components capable of being integrated into planar multilayer structures.

Most commonly used ferrites are strontium hexaferrite M-type ($SrFe_{12}O_{19}$, SrM) and the material in the focus of this thesis, barium hexaferrite M-type ($BaFe_{12}O_{19}$, BaM). For readers unfamiliar with the abbreviations of chemical elements, a list of elements can be found at the end of this work on page 94. There are six types of hexaferrite, referred to as M, W, Y, Z, X, and U type. Composition diagrams and information on the crystal structures of these hexaferrite types can be found in [9]. Excellent reviews on soft and hard magnetic ferrite materials and their properties and applications can be found in [10] and [5]. Because of its promising material properties for the application in self-biased nonreciprocal microwave components, this work focuses on BaM only.

Units

For the sake of clarity, a comment about the used units is necessary. Most practical works in the field of ferrites are performed using the *Centimetre Gram Second* (CGS) units instead of the *Système international d'unités (International System of Units)* (SI) units.

[4]Metamagnetics Inc., Advanced Ferrite Based Solutions, Ka-Band Self-Biased Circulators and Isolators, 2017.

Table 2.1: Magnetic units in the CGS and the SI unit system.

Quantity	Symbol	CGS unit	Conversion	SI unit
Magnetic flux density	B	Gauss (G)	10^{-4}	Tesla (T), $\frac{\mathrm{Wb}}{\mathrm{m}^2}$
Magnetic field strength	H	Oersted (Oe)	$\frac{10^3}{4\pi}$	$\frac{\mathrm{A}}{\mathrm{m}}$
Magnetic moment	m	emu	10^{-3}	Am^2, $\frac{\mathrm{J}}{\mathrm{T}}$
(Volume) magnetization	M	$\frac{\mathrm{emu}}{\mathrm{cm}^3}$	$\frac{10^3}{4\pi}$	$\frac{\mathrm{A}}{\mathrm{m}}$
(Mass) magnetization	σ	$\frac{\mathrm{emu}}{\mathrm{g}}$	1	$\frac{\mathrm{Am}^2}{\mathrm{kg}}$
Magnetic polarization	J	$\frac{\mathrm{emu}}{\mathrm{g}}$	$\frac{4\pi}{10^{-4}}$	T, $\frac{\mathrm{Wb}}{\mathrm{m}^2}$
Permeability	μ_0	1	$4\pi \cdot 10^{-7}$	$\frac{\mathrm{H}}{\mathrm{m}^2}$, $\frac{\mathrm{Wb}}{\mathrm{Am}}$

Table 2.1 list the units used in both systems and the corresponding conversion factors. The conversion factor is meant to be multiplied to the quantity in CGS units to get the value in SI units.

In SI units the relationship between $\vec{H}$, $\vec{B}$, and $\vec{M}$ is

$$\vec{B} = \mu_0(\vec{H} + \vec{M}), \tag{2.2}$$

whereas in CGS units, μ_0 is unity and therefore implies that $\vec{H}$ and $\vec{B}$ have the same numerical value in nonmagnetic materials. The corresponding equation is

$$\vec{B} = \vec{H} + 4\pi\vec{M}. \tag{2.3}$$

In CGS units, the Larmor frequency is

$$f_0 = \frac{\mu_0\gamma|\vec{H}_0|}{2\pi} = 2.8\,\frac{\mathrm{MHz}}{\mathrm{Oe}}\cdot|\vec{H}_0|, \tag{2.4}$$

whereas in SI units it is

$$f_0 = 28\,\frac{\mathrm{GHz}}{\mathrm{T}} \cdot |\vec{B}_0|. \tag{2.5}$$

In practice, these units are convenient and easy to use [7, p. 458]. As many publications in this field use CGS units, the CGS system is also used in this work.

2.1 Barium Hexaferrite

BaM is a hard magnetic ferrite material with a hexagonal crystal structure. It possesses the structure of magnetoplumbite $M_eO{\cdot}6Fe_2O_3$, in which M_e denotes one of the divalent ions: Ba^{2+}, Sr^{2+}, or Pb^{2+}. The structure was first described in [11] and material was developed in the Philips Research Laboratories [9]. It was first named ferroxdure in accordance with the cubic ferrite with spinel structure named ferroxcube.

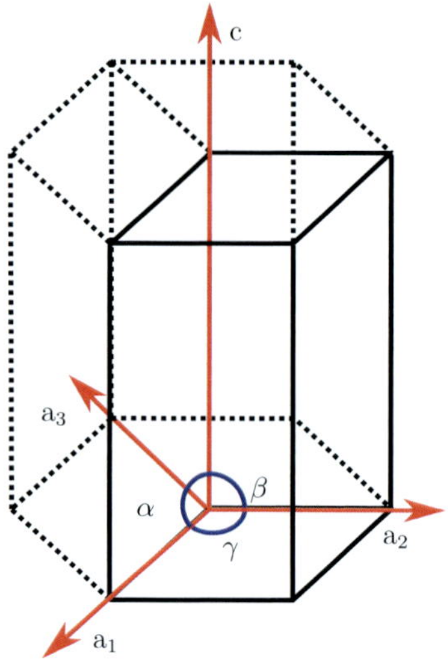

Figure 2.1: Hexagonal unit cell.

Figure 2.1 shows the hexagonal unit cell of BaM marked with solid lines. The unit cell is the smallest repeating structure of atoms, from which the structure of the complete crystal can be composed. The lengths along the a_i- and c-axis of the BaM unit cell are 5.89 Å and 23.2 Å, respectively [12]. The angles between the axes are $\alpha = \beta = 90°$ and $\gamma = 120°$.

Figure 2.2 shows a perspective view of a unit cell with barium, oxygen, and iron ions. It can be seen that the cell consists of two fundamental building blocks: R and S. The blocks R^* and S^* are derived by rotating the R and S blocks by 180° about the c-axis. The R^* and S^* blocks are stacked in alternation with the R and S blocks. A unit cell is built from four

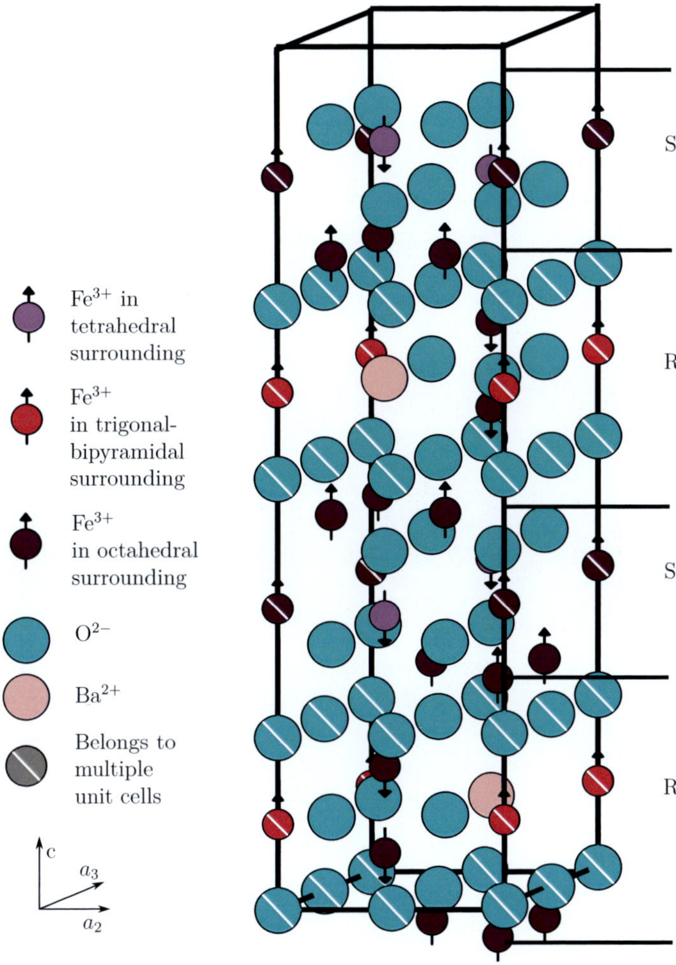

Figure 2.2: Perspective representation of the magnetoplumbite structure with the c-axis vertical. The arrows indicate spin orientation.

stacked blocks along the c-axis. Thus, ten layers of oxygen anions are stacked along the c-axis and each R and R^* block contains one Ba^{2+} cation. Therefore, a unit cell contains 64 ions, namely 38 oxygen ions, 24 iron ions, and two barium ions ($Ba_2Fe_{24}O_{38}$). The trivalent iron ions occupy octahedral, tetrahedral and, in the barium layer, trigonal-bipyramidal sites. Their spins are aligned in two antiparallel sublattices and the orientation for each Fe^{3+} ion is indicated by an arrow in Figure 2.2. Here, an upwards/downwards-directed arrow indicates a contribution to the majority/minority sublattice. The difference between the magnetic moments of the two sublattices determines the saturation magnetization of the material. In a unit cell, the major sublattice contains eight more iron ions than the minor sublattice, each Fe^{3+} ion contributing five μ_B (bohr magneton). This yields a theoretical saturation magnetization M_s of 6632 G at 0 °K and 4775 G at 20 °C [6].

BaM has a strong uniaxial magnetocrystalline anisotropy with the preferred direction of magnetization, the so-called easy axis of magnetization, along the c-axis. The primary origin of the magnetocrystalline anisotropy is spin-orbit interaction, which is the interaction between the magnetic field generated by the electron orbiting the nucleus and the spin of the electron. Another cause of the crystalline anisotropy is dipole-dipole interaction. Further information on both mechanisms can be found in [6].

The anisotropy field H_A of BaM is close to 17 kOe [13]. According to (2.4), this field yields a Larmor frequency close to 47.6 GHz. Thus, the material is suitable for applications in the Q-/V-band without an externally applied biasing field. Section 3.4 describes how the anisotropy of the material is influenced by substituting iron ions in different positions for Sc or Al ions. This is done in order to vary the anisotropy field and thus make the material usable over a broad range of frequencies.

2.2 Interaction with Electromagnetic Fields

The following gives a brief introduction to the interaction of ferrite materials and electromagnetic fields. A more elaborate and thorough introduction can be found in [7]. In general, there is interaction between microwave fields and the spinning electrons in all ferromagnetic and ferrimagnetic materials, but it is only in electrically-insulating materials

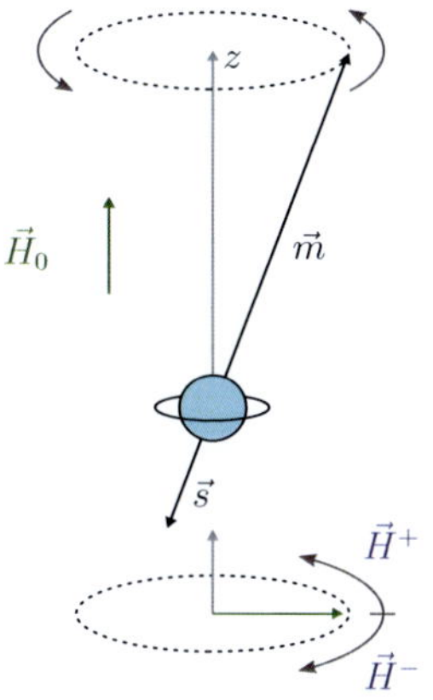

Figure 2.3: Magnetic dipole moment and angular momentum of a spinning electron.

such as ferrites that useful interaction for nonreciprocal microwave components can be obtained.

Figure 2.3 shows the commonly used analogy of a spinning electron with an angular momentum $\vec{s}$ and a magnetic moment $\vec{m}$ as indicated. A constant magnetic field $\vec{H}_0$ is applied along the z-axis and positive and negative circularly-polarized fields $\vec{H}^+$ and $\vec{H}^-$, which will be of practical use in Section 3.3, are indicated. The relation between $\vec{m}$ and $\vec{s}$ is given by

$$\vec{m} = -\gamma\vec{s} \tag{2.6}$$

with $\gamma = 1.758 \cdot 10^{11}\,\mathrm{C} \cdot \mathrm{kg}^{-1}$ the gyromagnetic ratio. Due to the magnetic bias field $\vec{H}_0$, a torque

$$\vec{T} = \mu_0 \vec{m} \times \vec{H}_0 = \frac{d\vec{s}}{dt} = \frac{-1}{\gamma}\frac{d\vec{m}}{dt} \tag{2.7}$$

is exerted on the magnetic dipole with the time rate of change of the angular momentum being equal to the torque. This yields the equation of motion

$$\frac{d\vec{m}}{dt} = -\mu_0\gamma\vec{m} \times \vec{H}_0. \tag{2.8}$$

The solution of this equation entails that $\vec{m}$ precesses about the z-axis with the precession frequency

$$\omega_0 = \mu_0 \gamma |\vec{H}_0|, \tag{2.9}$$

the Larmor frequency mentioned previously. The total magnetization $\vec{M}_\mathrm{t}$ of a material is the sum of all N unbalanced magnetic dipoles in a unit volume:

$$\vec{M}_\mathrm{t} = N\vec{m}. \tag{2.10}$$

As applied in Section 2.1, this equation can be used to calculate the saturation magnetization M_s of the BaM unit cell. If an additional microwave field $\vec{H}$ is applied, the total magnetization, assuming the material is magnetized to saturation, is

$$\vec{M}_\mathrm{t} = M_\mathrm{s}\vec{u}_z + \vec{M}, \tag{2.11}$$

with M_s in direction of the z-axis and $\vec{M}$ being the additional magnetization due to $\vec{H}$. With the assumption that $\vec{H} \ll \vec{H}_0$, products of $\vec{M}$ and $\vec{H}$ can be neglected. Using (2.9) and (2.10), (2.8) yields the equations of motion

$$\frac{d^2 M_x}{dt^2} + \omega_0^2 M_x = \omega_\mathrm{m}\frac{dH_y}{dt} + \omega_0\omega_\mathrm{m} H_x \tag{2.12}$$

and

$$\frac{d^2 M_y}{dt^2} + \omega_0^2 M_y = -\omega_\mathrm{m}\frac{dH_x}{dt} + \omega_0\omega_\mathrm{m} H_y, \tag{2.13}$$

with $\omega_\mathrm{m} = \mu_0\gamma M_\mathrm{s}$ and subscripts x and y denoting the components of the magnetization and the magnetic field. Assuming that $\vec{H}$ has a time-harmonic dependence according to $e^{j\omega t}$, these equations yield

$$(\omega_0^2 - \omega^2)M_x = \omega_0\omega_\mathrm{m} H_x + j\omega\omega_\mathrm{m} H_y \tag{2.14}$$

and

$$(\omega_0^2 - \omega^2)M_y = -j\omega\omega_\mathrm{m} H_x + \omega_0\omega_\mathrm{m} H_y\,. \tag{2.15}$$

Therefore, $\vec{H}$ and $\vec{M}$ are related by the susceptibility tensor $[\chi]$

$$\vec{M} = [\chi]\vec{H} = \begin{bmatrix} \chi_{xx} & \chi_{xy} & 0 \\ \chi_{yx} & \chi_{yy} & 0 \\ 0 & 0 & 0 \end{bmatrix} \vec{H}. \tag{2.16}$$

The elements of $[\chi]$ are

$$\chi_{xx} = \chi_{yy} = \frac{\omega_0 \omega_\mathrm{m}}{\omega_0^2 - \omega^2} \tag{2.17}$$

and

$$\chi_{xy} = -\chi_{yx} = \frac{j\omega\omega_\mathrm{m}}{\omega_0^2 - \omega^2}. \tag{2.18}$$

$\vec{B}$ and $\vec{H}$ are related by

$$\vec{B} = \mu_0(\vec{M} + \vec{H}) = [\mu]\vec{H} \tag{2.19}$$

with the permeability tensor

$$[\mu] = \begin{bmatrix} \mu & j\kappa & 0 \\ -j\kappa & \mu & 0 \\ 0 & 0 & \mu_0 \end{bmatrix}. \tag{2.20}$$

The elements of the tensor are

$$\mu = \mu_0(1 + \chi_{xx}) = \mu_0 \left(1 + \frac{\omega_0 \omega_\mathrm{m}}{\omega_0^2 - \omega^2}\right) \tag{2.21}$$

and

$$\kappa = -j\mu_0 \chi_{xy} = \mu_0 \frac{j\omega\omega_\mathrm{m}}{\omega_0^2 - \omega^2}. \tag{2.22}$$

The permeability tensor is also called Polder tensor [14] and is valid for ferrite materials which are magnetized to saturation.

According to (2.9), the Larmor frequency, at which the ferromagnetic resonance (FMR) occurs is determined by the internal, static, magnetic bias field $\vec{H}_0$. $\vec{H}_0$ is in turn determined by an externally applied bias field $\vec{H}_\mathrm{E}$, demagnetizing fields, and the anisotropy field $\vec{H}_\mathrm{A}$, which is assumed to be parallel to $\vec{H}_\mathrm{E}$. The FMRF f_r of the ferrite material with the bias field applied along the z-axis is given by [15]

$$f_\mathrm{r} = \frac{\mu_0 \gamma}{2\pi} \sqrt{(H_\mathrm{E} + H_\mathrm{A} + (N_x - N_z)M_\mathrm{s})(H_\mathrm{E} + H_\mathrm{A} + (N_y - N_z)M_\mathrm{s})}. \tag{2.23}$$

N_x, N_y, and N_z are the shape-dependent, and therefore direction-dependent, demagnetization factors. These factors take into account the difference between internal and external magnetic field due to the boundary conditions at the ferrite surface. In this work, thin ferrite films are prepared and for this simple shape the demagnetization factors are $N_x = 1$ and $N_y = N_z = 0$, with the x-axis being perpendicular to the film plane. The externally applied bias field $\vec{H}_\mathrm{E}$ for self-biased ferrites is 0. For a ferrite film with in-plane (IP) anisotropy and magnetization M this yields

$$f_\mathrm{r} = \frac{\mu_0 \gamma}{2\pi} \sqrt{(H_\mathrm{A} + M) H_\mathrm{A}}. \tag{2.24}$$

The magnetocrystalline anisotropy field therefore determines the FMRF of self-biased ferrite materials which is thus controlled by the choice of the ferrite material.

2.3 Material Synthesis and Integration

There are various methods that can be used for the synthesis of BaM films. In the following Section 2.3.1 a brief overview is given. A more detailed overview including methods to produce powder and bulk material can be found in [16]. The sol-gel method, which is used in this work to produce films for the integration into LTCC, is described in more detail in Section 2.3.2 and further information can be found in [17] and [4]. There are several approaches for the integration of ferrite materials into LTCC. Advantages and disadvantages of these techniques are discussed in Section 2.3.3. In this work, films are fabricated which can be applied in a post-firing step.

2.3.1 Synthesis of Barium Hexaferrite Films

In [18], the deposition of BaM films on alumina substrates by magnetron sputtering is described. The films have a thickness of 10-15 µm and are produced with a sputtering rate of $0.75\,\mu\mathrm{m} \cdot \mathrm{h}^{-1}$. The used sintering temperature of 800 °C is rather low compared to conventional ceramic methods with sintering temperatures of up to 1400 °C [16]. The disadvantages this method presents are that the produced films are not textured and that rather expensive equipment is required. As will be discussed in Section 3.5, it is advantageous to produce textured films, to improve the magnetic properties of the material. A circulator using sputter-deposited BaM films is presented in [19].

BaM films grown by chemical vapour deposition (CVD) are presented in [20]. In [21], CVD is used to synthesize BaM films with a thickness of 270 nm ($1.4\,\mu m \cdot h^{-1}$ growth rate) on basal plane sapphire and thermally-oxidized Si (Si/SiO_2) substrates. The temperature in the reaction zone is 800 °C and the films produced on the sapphire substrate exhibit a strong *c*-axis orientation perpendicular to the substrate plane, while the films produced on thermally-oxidized silicon are randomly oriented. The advantages this method offers are the low-reaction temperature and the produced texture. However, the orientation of the *c*-axis depends on the used substrate material and expensive equipment is required.

In [22], pulsed laser deposition (PLD) is used to produce BaM films on basal plane sapphire. The thickness of the produced films is 440 nm. The surface temperature of the substrate during film deposition is 800 °C and post annealing in an oxygen atmosphere is performed at temperatures from 900 °C to 1000 °C. The films show a preferred orientation with the *c*-axis out-of-plane (OP) and barium monoferrite ($BaFe_2O_4$) is detected as a secondary phase. Compared to conventional ceramic methods, the post-firing temperature of up to 1000 °C in PLD is rather low, but it is high in comparison with the other methods discussed here.

BaM films prepared by liquid-phase epitaxy (LPE) are presented in [23]. The films are grown on sapphire substrates with the *c*-axis IP. In a first step, PLD is used to apply a seed layer and in a second step a BaM film with a thickness from 60 µm to 200 µm is grown by LPE. In [23], no temperature for the film growth is given, but, for a similar method in [24], a temperature of 860 °C is used after a mixing step at 1250 °C. The LPE method is capable of producing BaM thick films of high quality at rather low temperatures, but it requires expensive equipment and suitable substrate material.

In [25] and [26], BaM films are prepared with the metallo-organic decomposition (MOD) method on a Si/SiO_2 wafer using a sputter-deposited Pt interlayer. The films, which are spin coated in three to six steps, have an overall thickness of up to 460 nm and are deposited on a Pt layer with a thickness of 300 nm. Various tempering steps in different atmospheres are used in the fabrication with the highest temperature being 900 °C in an O_2 atmosphere. The films show a strong OP *c*-axis orientation. The MOD method is capable of producing thin, textured BaM films, but the texture depends on the substrate material and the Pt interlayer. Especially, this Pt interlayer limits the applications for these films.

In [27], screen-printed BaM films with a thicknesses ranging from 200 μm to 500 μm are presented. A barium hexaferrite powder is prepared by conventional ceramic processing and is used in a screen-printing paste. The printed films are dried in a magnetic field to align the *c*-axis of the BaM particles at temperatures ranging from 150 °C to 250 °C. In a second sintering step at temperatures from 900 °C to 1300 °C, the film is hot-pressed to improve the density. Advantages of this method are the high film thickness and the texture of the produced film. However, in comparison to other methods it requires a rather high sintering temperature. An approach, which uses a similar technique to produce a textured material, is used in [28] to produce a composite magnetic material. BaM nanoparticles are dispersed in photoresist and the mixture is spin-coated on alumina substrates. Similar to the screen-printing approach, the particles are aligned by a magnetic field while the photoresist is dried. In [28], a circulator using this kind of composite material is presented. Due to the focus of the publication on the circulator, no material analysis results are presented. Also, the temperature of the post-exposure bake is not stated.

The BaM films in this work are produced using the sol-gel method. In comparison to the other methods, the sol-gel method is rather simple and low-cost. Only readily available chemicals are needed as starting materials in comparison to sputtering, PLD and screen-printing where BaM is needed either as target or as powder. Next to standard laboratory equipment, a spin-coater and an oven capable of providing the sintering temperature are needed. Another major advantage is a potentially low sintering temperature and the fact that there is no need for sintering in controlled atmospheres. In comparison to the other methods discussed, the equipment needed to produce BaM films using the sol-gel method is quite affordable. However, achieving a high film thicknesses and preferred orientation of the particles on a given substrate material is challenging.

In Table 2.2, a brief overview of the production methods discussed above is given. Crucial for the compatibility with LTCC substrates is a maximum processing temperature of 850 °C [29]. Furthermore, a high film thickness is preferable and it is important to notice that for most methods the film texture depends on the structure of the used substrate material. The column 'Equipment' is meant as an indication of expense and is a strikingly simplified comparison.

Table 2.2: Comparison of BaM film fabrication methods.

Method	Max. temp.	Film thickness	Substrate	Texture	Equipment
Magnetron sputtering [18]	800 °C	10 – 15 µm	Alumina	-	Extensive
CVD [20], [21]	800 °C	270 nm	Sapphire	OP	Extensive
PLD [22]	900 °C - 1000 °C	440 nm	Sapphire	OP	Extensive
LPE [23], [24]	870 °C	200 µm	Sapphire	IP	Extensive
MOD [25], [26]	900 °C	460 nm	$Si/SiO_2/Pt$	OP	Extensive
Screen printing [27]	900 °C - 1300 °C	200 µm - 500 µm	Alumina	✓	Simple
Sol-gel	850 °C	10 µm	Alumina/ LTCC	-	Simple

2.3.2 Film Preparation by Means of the Sol-Gel Method

The sol-gel method used in this work has been previously applied to produce BaM films for different applications. In the following, some work on this topic is briefly summarized, focusing on the used substrate materials, achieved film thicknesses, and major differences in the preparation processes used.

The fabrication of BaM films by means of the sol-gel method is presented in [30]. The films are prepared on thermally-oxidized silicon substrates and have a thickness of up to 300 nm. Formation of BaM is observed at temperatures above 600 °C. The production of Barium hexaferrite powder by means of the sol-gel method is described in [31] and preheating is discussed as a key step in avoiding the formation of hematite as a secondary phase. The effect of vacuum extraction while homogenizing the precursors and two compositions of the precursor solution are discussed in [32]. In [33] and [34], the preparation of barium hexaferrite films on an Si wafer with a thermally-oxidized SiO_2 layer by means of a water-based sol-gel process is demonstrated. The derived films have a thickness of 120 nm and are sintered at a low temperature of 650 °C. In [35], nanocrystalline barium hexaferrite films with a crystallite size of 23 nm and a film thickness from 150 nm to 160 nm are presented. The films are produced on Si substrates and the crystallite size is influenced by several parameters in the fabrication process. Films with high Fe/Ba ratios in the precursor solution are analyzed in [36]. The films are prepared on Si and have a thickness of 130 nm. The results show that the film with the lowest tested ratio of 6.5 exhibits the best magnetic properties in comparison to films with ratios of up to 12. In [37], films with a thickness of 0.8 µm are prepared on sapphire substrate with the films having the *c*-axis orientation of the substrate. The effect of ethylene glycol in the precursor solution on the orientation of films produced on oriented sapphire substrates is investigated in [38]. The Fe/Ba ratio used in that work is 6, the thickness of the produced films is 150 nm, and rather high sintering temperatures from 900 °C to 1035 °C are used.

The films in this work are produced on polycrystalline alumina (99.6% corundum, Al_2O_3) substrates with no texture by Cicor Advanced Microelectronics & Substrates. The used sol-gel process is described in the following. Aims are a high film thickness, a low sintering temperature, and good magnetic properties. The influence of several process parameters on the produced film is investigated and discussed in Section 3.

Figure 2.4 shows the steps used in this work to produce BaM films. The precursor materials barium nitrate ($Ba(NO_3)_2$), iron nitrate ($Fe(NO_3)_3 \cdot 9H_2O$), and, for the doped films presented in Section 3.4, scandium nitrate ($Sc(NO_3)_3$) or aluminum nitrate ($Al(NO_3)_3 \cdot 9H_2O$), are dissolved in ethylene glycol ($HOCH_2CH_2OH$). The weight of the precursors is calculated from the desired ratio of the elements in the solution and measured using an analytical scale. The weighted precursor materials are dissolved in ethylene glycol, with the solution containing 80 % weight ethylene glycol.

The aim of the next step is to dissolve the powder precursors completely and to homogenize the solution. In a first attempt ultrasonic stirring was used, but it did not dissolve the precursor materials completely before gelation of the solution. Because of the required open setup, the solution may evaporate during homogenization. This step was optimized at the beginning of this work.

As an alternative approach, the solution is refluxed as depicted in Figure 2.5. The solution is heated in a flat bottom flask on a hot plate and a water bath is used to keep the temperature constant. Due to the reflux cooler, the solution can be homogenized for 24 h at 70 °C without drying. During this time, a magnetic stir bar rotates with 500 rpm.

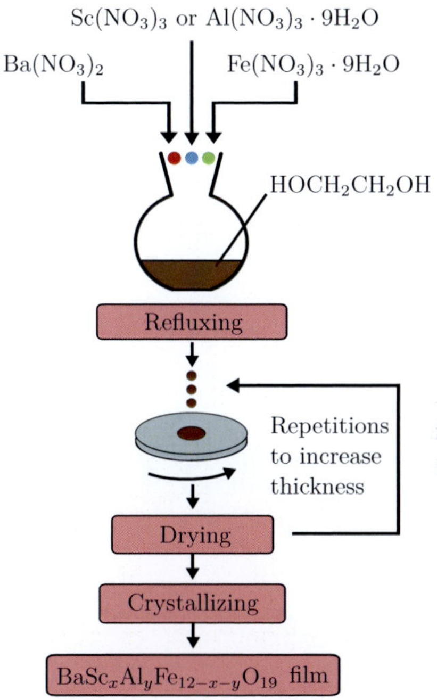

Figure 2.4: Steps to produce barium hexaferrite films by means of the sol-gel method.

Using this approach, the precursor materials are dissolved completely and a homogeneous solution is obtained.

In the next step, a thin film is deposited on the substrate by spin coating. The substrates are cleaned with acetone and isopropanol before the film is deposited. The employed spin coater allows for adjusting the number of revolutions per minute up to 4000 rpm. The spin time was chosen to be 20 s. In the next step, the produced layer is dried in the laboratory oven B180 by Nabertherm. The optimum drying time and temperature are discussed in Section 3.1. The deposition and drying steps are repeated several times to deposit more layers and therefore achieve a higher final film thickness.

In the last step, the film is sintered in the laboratory oven without a synthetic atmosphere. The sample is heated from the drying temperature to the sintering temperature with 10 °C

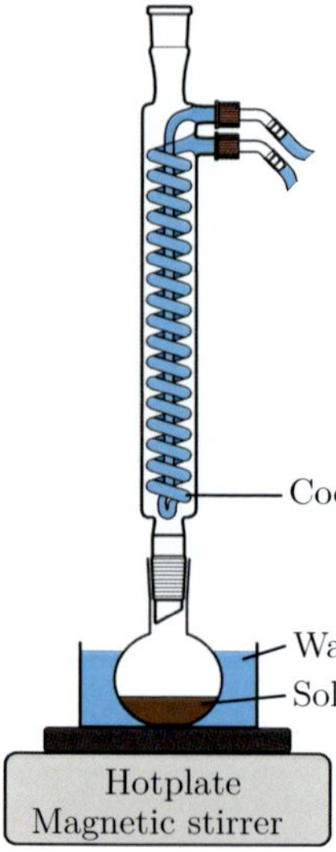

Figure 2.5: Homogenization of the precursor solution using a hotplate with water bath and refluxing cooler.

per min. The influence of sintering temperature and sintering time on the produced film is discussed in Section 3.2.2. At the end of the sintering process, the oven is turned off and the sample is slowly cooled to room temperature inside the closed oven. Due to the slow natural cooling, thermal stress in the produced film is avoided.

For certain applications, it is necessary to have a ferrite film only in specified areas of the substrate, not on the whole surface. If the films are produced using a low drying temperature, either lift-off processes can be used or, in simple cases, some areas of the surface can be protected using kapton tape. This does not work if high drying temperatures are used. Kapton is unstable at temperatures above 400 °C and the adhesive of the tape burns at even lower temperatures. Photoresists cannot be removed easily after being exposed to temperatures of roughly more than 200 °C, with the exact temperature depending on the used photoresist. In another approach, a flat high temperature gasket, kindly supplied by Hendricks-Indutec, in combination with a vacuum chuck built to fit and apply pressure to the substrate was used to cover certain substrate areas while spin coating. The high pressure that is needed to seal the gasket in order to avoid contamination with the coating material leads to cracking of the alumina substrates. Therefore, the gasket did not solve this problem.

As an alternative to the above outlined approaches, the sample surface is coated completely, and after sintering, the ferrite is etched to yield the desired shape. For this, a layer of photoresist MA-P1215 from Micro Resist Technology is deposited on the sintered ferrite film by spin coating. After exposure and developing of the photoresist, the ferrite can be etched with heated hydrochloric acid (HCl). A problem occurring in this step is that gold conductors can be peeled off the substrate during etching of the ferrite. It is challenging to etch the ferrite completely without damaging the conductors. For a component relying on these films, it would therefore be desirable to avoid structuring of the ferrite if a metalization beneath the ferrite film is required.

2.3.3 Ferrite Integration into LTCC

The different sintering temperatures of the methods outlined above are of importance for the integration of the films into LTCC. LTCC technology offers multiple advantages such as lower losses at high frequencies compared to organic materials, a low thermal expansion coefficient, the possibility of building various types of components by combining materials with different characteristics in a single multilayer device [3]. In contrast to high temperature co-fired ceramics, low-loss metals can be incorporated. LTCC substrates are therefore an attractive choice for the design of small and lightweight components working in the Q-/V-band for future applications in satellite communications. The integration of ferrite materials capable of self-biasing would enable the design and integration of compact nonreciprocal components.

Figure 2.6 depicts the steps required for the LTCC fabrication. A so-called green tape is the starting material. In a first step, the tape is blanked, vias are punched into the tape and filled with a conductive paste. Afterwards, the conductors are screen-printed with a metal paste, the separate layers are stacked, and laminated. In the co-firing step, the ceramic and the integrated conductors are sintered at temperatures typically close to 850 °C, depending on the chosen LTCC material. A specific sintering profile for each substrate is specified by the corresponding manufacturer. After co-firing, the ferrite film can be deposited, possibly

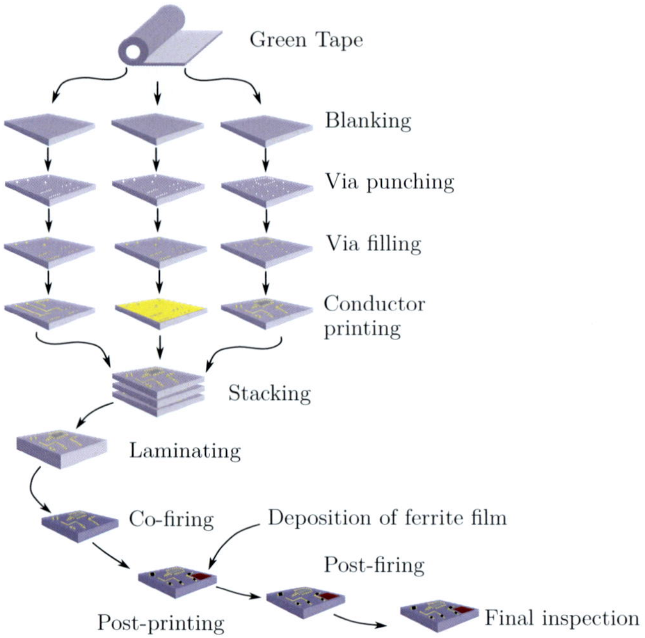

Figure 2.6: Overview of the LTCC fabrication process with deposition of the ferrite film after co-firing.

along with other materials such as resistive pastes. In a post-firing step, the ferrite film is sintered. The temperature in this step is critical because reactions of the film with the substrate material and diffusion at the interface of film and substrate can occur at high temperatures. Furthermore, the LTCC material changes its dielectric properties if high temperatures are used in post-firing steps [29].

LTCC is not suited for methods of producing textured films that depend on a substrate with a certain crystal structure and orientation to obtain the c-axis alignment. Examples of such methods are LPE, CVD, PLD, and MOD, which are discussed in Section 2.3.1. Only the screen-printing method and the magnetic composite material, in which the preferred orientation is achieved by applying a magnetic field to the film while drying, work rather independently of the substrate material. Therefore, these methods could produce textured films on LTCC substrates.

There are other approaches to integrate ferrite materials into LTCC. In [39] and [40], a soft magnetic ferrite tape is co-fired with the dielectric tapes. Challenges this approach presents are finding a ferrite tape with low sintering temperature and suitable shrinkage rates to avoid cracks and warpage. Since it is difficult to produce such a material with acceptable microwave properties, alternative approaches are of interest. In [41], a soft magnetic pre-sintered ferrite pill is integrated into a cavity in the LTCC. For this approach, the exact determination of the LTCC shrinkage is crucial. To avoid stress and gaps after sintering, the size of the cavity has to be chosen carefully. Both methods use soft magnetic ferrites and therefore need permanent magnets to supply a biasing field. However, both could possibly also work with hard magnetic ferrite materials. Tapes of hard magnetic ferrite are not available yet, although producing a suitable ferrite pill is feasible.

2.4 Material Analysis

In this work, several methods of material analysis are applied, to determine the magnetic and structural properties of the films. The used techniques and the required sample preparations are described in the following. Furthermore, S-parameter measurements, which are described in Section 3.3, are performed in this work.

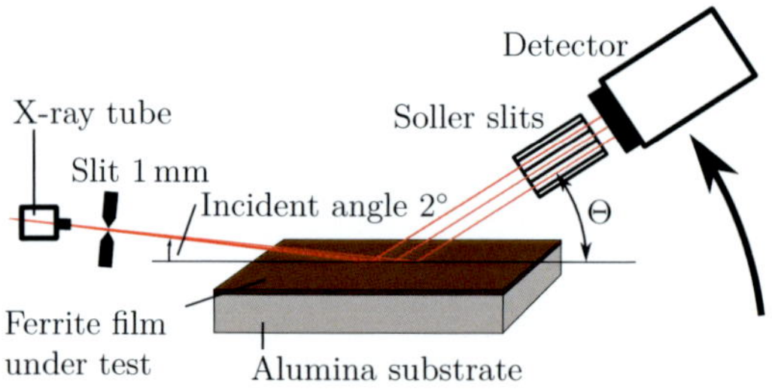

Figure 2.7: Schematic view of an XRD-analysis performed using the grazing incident method.

2.4.1 X-Ray Diffraction

To analyze the produced material and investigate the formation of secondary phases, X-ray diffraction (XRD)-measurements are performed. The XRD-analyses in this work are performed at the Institute of Advanced Ceramics and the Institute of Materials Physics and Technology, both part of the Hamburg University of Technology (TUHH). A Bruker AXS D8 diffractometer system with Cu K-α radiation was used.

The XRD analysis is based upon the diffraction of monochromatic X-rays on a crystalline sample as depicted in Figure 2.7. It relies on Bragg's law

$$\lambda = 2d_{hkil} \cdot \sin(\Theta), \tag{2.25}$$

where d_{hkil} is the spacing of a set of crystallographic planes with the Miller-Bravais indices $hkil$. The wavelength λ of the used Cu K-α radiation is 1.54 Å and Θ is the Bragg angle. One of these planes passes through the origin and the next, as depicted for one exemplary plane in Figure 2.8, makes intercepts with the crystallographic axes a_1, a_2, a_3, and c at the points h, k, i, and l, respectively. In Figure 2.8, these intercepts for the exemplary plane are at a, ∞, -a, and c and, except for ∞, they are marked by green points. These intercepts h, k, i, and l are normalized to the unit cell dimensions given in Section 2.1, yielding Miller-Bravais indices $(10\bar{1}1)$ for the exemplary plane.

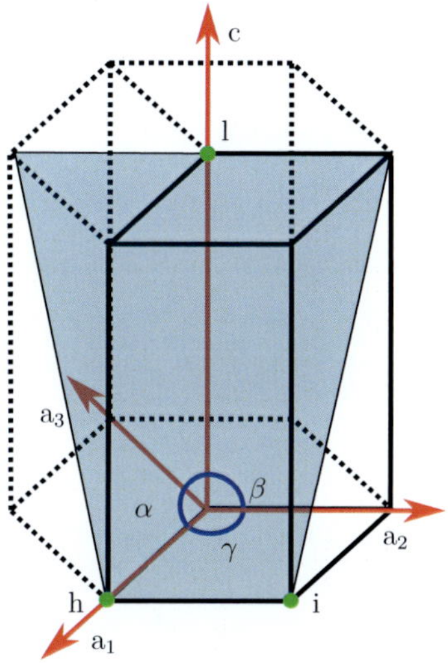

Figure 2.8: Hexagonal unit cell with an exemplary plane for XRD-analysis indicated in blue.

Further information can be found in [42]. The diffracted X-rays are detected over the angle range of interest and the characteristic diffraction peaks are recorded. The films produced in this work consist of small randomly-oriented crystallites, therefore results similar to those of a powder sample are expected. The observed position and intensity of the diffraction peaks are used for a structural analysis. The crystalline phases in the material can be identified by their distinctive diffraction pattern. The lattice parameters can be calculated using (2.25), since each peak represents certain lattice planes with distances d_{hkil}. In this work reference patterns obtained from [43]–[45] are used. Furthermore, the texture of the sample can be analyzed by matching the peaks to the planes and comparing the observed intensities to those of an untextured sample. Also, the peak width can give an indication for strain in the material and the size of the observed crystallites.

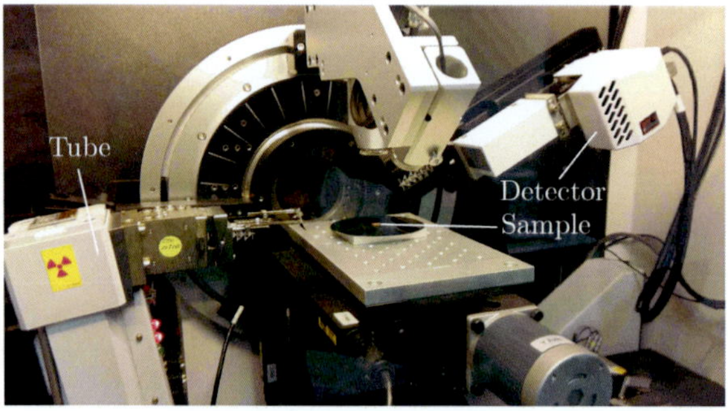

Figure 2.9: Picture of the performed XRD-analysis.

The results for the films are obtained using the grazing incidence diffraction (GID) method as depicted in Figure 2.7. In this technique, the tube is permanently pointed at a very low angle, in this case 2°, towards the substrate and a detector scan is performed. The GID analysis is highly receptive at the sample surface due to the low angle of the X-ray source and therefore well suited for the analysis of thin films.

Figure 2.9 shows a photograph of the measurement setup. The analyzed sample can be seen in the center of the picture on a zero background wafer. The sample films for the XRD-analyses are prepared on square alumina substrates with an edge length of 0.85 mm. To avoid the contamination of the substrate surface with remaining gold or wash primer, the substrates are thoroughly cleaned before film deposition. The mentioned residues would otherwise result in additional diffraction peaks which could complicate the interpretation of the results and possibly mask other peaks.

2.4.2 Scanning Electron Microscopy

To produce images with a high magnification of the film surface and cross section, a Zeiss Supra 55 VP scanning electron microscope (SEM) is used in this work. In SEM, the sample is scanned in a raster scan pattern with a highly focused electron beam. The secondary and backscattered electrons are detected using an in-lens and a secondary-electron detector, creating an image of the sample surface. To avoid charging of the sample surface, a low acceleration voltage of 3 kV–5 kV and a rather small apertures of 7.5 µm–10 µm are used.

Furthermore, the SEM can be used to obtain information about the stoichiometry of the material. The high energy electron beam interacts with the atoms on the sample surface and excites electrons on the inner shells and ejects them. The newly formed holes are then filled by electrons from an outer shell. Due to the energy difference between the two shells, a characteristic X-ray is emitted. Using an X-ray detector inside the specimen chamber for an energy-dispersive X-ray spectroscopy (EDXS) analysis, these rays can be detected, the atoms on the surface can be identified, and the stoichiometry of the sample can be determined. Since the electron beam is highly focused, maps of the different atoms on the sample surface can be produced. To avoid surface charges due to the high energy electron beam, a ~10 nm gold layer is sputter-deposited on the sample prior to the EDXS measurements. All SEM samples are attached to the aluminum sample holder using adhesive, conductive tabs. Other than that, there are no specific requirements for the SEM and EDXS samples. The sample preparation and analysis are performed at the electron microscopy unit of the TUHH.

2.4.3 VSM-Measurement

The magnetic hysteresis loop of the produced materials is obtained by vibrating sample magnetometer (VSM)-measurements, which are performed using a Quantum Design vibrating sample magnetometer PPMS VersaLab at the Institute of Nanostructure and Solid State Physics of the University of Hamburg (UHH).

In a VSM-measurement, the sample is magnetized by a magnetic field and then vibrated to induce a voltage in pickup coils. The induced voltage is proportional to the magnetic moment of the sample, thus making it possible to record the hysteresis loop of the material. In this work, hysteresis loops are recorded with magnetic fields from −30 kOe to 30 kOe. The sample size for these measurements is restricted to substrates with edge lengths of less than 3 mm. Since the spin-coated films are uneven at the substrate edges, it is not possible to deposit the films in a reproducible manner on such small substrates. To produce these small film samples, larger substrates are slightly slitted on the backside prior to the

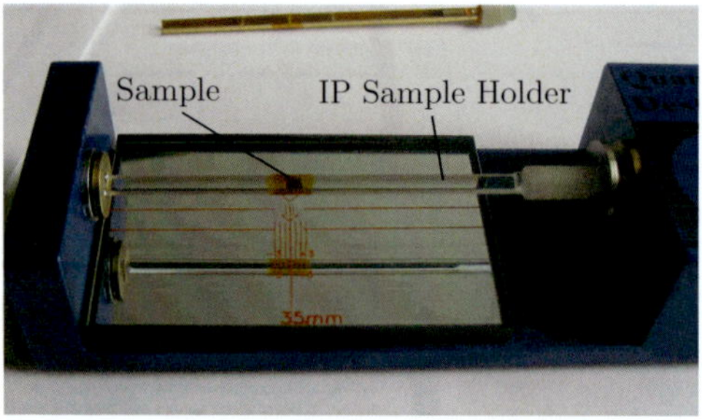

Figure 2.10: Sample for VSM-measurements attached to IP quartz sample holder using kapton tape. The sample holder is positioned in an alignment aid.

deposition of the film. After sintering, the substrates are broken along these slits. This way, small samples with a homogeneous film surface can be obtained.

Figure 2.10 shows a sample, which is attached with kapton tape to the quartz panel sample holder for IP measurements of the hysteresis loop. OP measurements are performed using a tube sample holder.

In this work, the sample volume is determined rather than the film mass, therefore the magnetization is not given in $emu \cdot g^{-1}$ but in $emu \cdot cm^{-3}$. This is done because determining the weight or density of the small film sample accurately is difficult. The surface area for each sample can be determined using a scanner with high resolution and an image-processing software. The film thickness can be determined from SEM images of the sample cross section.

2.4.4 Differential Scanning Calorimetry

To detect phase transitions in the material while heating it up, a differential scanning calorimetry (DSC) analysis is performed at the Insitute of Advanced Ceramics of TUHH. For this analysis, the material under test is filled into an alumina crucible. The filled crucible and an empty reference crucible are slowly heated over the temperature range of interest. The heat flux into both crucibles is monitored and therefore exothermic and endothermic reactions of the material under test can be detected. A thermogravimetric analysis (TGA) is performed simultaneously during heating of the material to monitor the weight loss of the sample during the heating process. For this, a Mettler Toledo TGA/DSC 1 is used. The sample preparation for the DSC and the TGA is described in detail in Section 3.2.2.

3 Synthesis and Properties of Barium Hexaferrite Films

In this chapter, the fabrication and the influence of several process parameters on the produced BaM film are discussed. Firstly, the production of crack-free films with a good adhesion to the substrate is described. In Section 3.2, the formation of secondary phases is analyzed and the fabrication of single-phase BaM films is outlined. Afterwards, microwave measurements without an externally applied bias field are presented and, in Section 3.4, the films are doped to influence the anisotropy field and thus the resonant frequency of the material. At the end of this chapter, the synthesis of BaM films on different LTCC materials is tested, analyzed, and discussed.

3.1 Film Surface and Adhesion to Substrate

As described in Section 2.3.2, the barium hexaferrite films are prepared by using the sol-gel method on alumina (Al_2O_3) substrates. This process is similar to the one in [30], where Barium hexaferrite films are produced on thermally-oxidized silicon substrates (SiO_2/Si). Utilizing similar process parameters for the production of films on alumina substrates results in cracks in the film surface and a poor adhesion of the produced material to the substrate. Exemplary samples illustrating these problems are shown in Figure 3.1.

Figure 3.1(a) shows a light microscopy image of a sample prepared with a rotation speed of the sample during spin coating of 50 rps, drying temperature T_D of 200 °C, drying time of ten minutes, and sintering temperature T_S of 800 °C. The solution was homogenized utilizing ultrasonic mixing at 80 °C and 15 layers were produced. The surface shows cracks and the film is partially peeling off the substrate. Figure 3.1(b) shows a film prepared on a

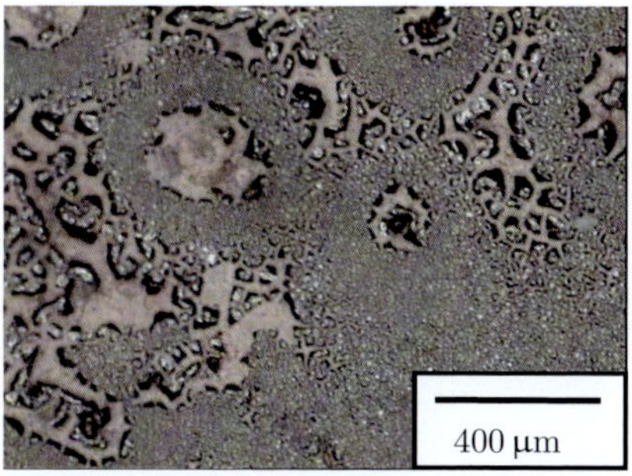

(a) Material produced with 50 rps on alumina substrate.

(b) Film prepared with 100 rps on alumina substrate with a gold conductor.

Figure 3.1: Samples prepared in an early stage of this work.

gold coplanar waveguide (CPW) on top of an alumina substrate. The material in the second photograph is produced with similar parameters except for a rotation speed of 100 rps and ten deposited layers. Both the higher rotation speed and the reduction in number of layers are used to lower the film thickness in an attempt to reduce the stress in the film during drying and sintering. The resulting material in Figure 3.1(b) shows cracks, delamination, and no significant improvement compared to the previous film. These examples show that changes in the preparation process of the film are necessary to synthesize a reproducible, homogeneous, and comparably thick film that is suitable for the fabrication of microwave components. This will be discussed in the following.

According to [4, p. 406] and [17, p. 259], thin films with a thickness below ~500 nm should dry without cracks. In an attempt to improve the results, several process parameters are varied and their influence on the film surface is analyzed. These experiments aim at reducing the layer and the film thickness and, ultimately, the stress by an adding polyvinylpyrrolidone (PVP) to the precursor solution. The reduced film thickness is achieved by using higher rotation speeds while spin coating and a lower viscosity of the solution. However, neither the higher rotation speed nor the addition of PVP to the solution lead to the formation of a crack-free surface.

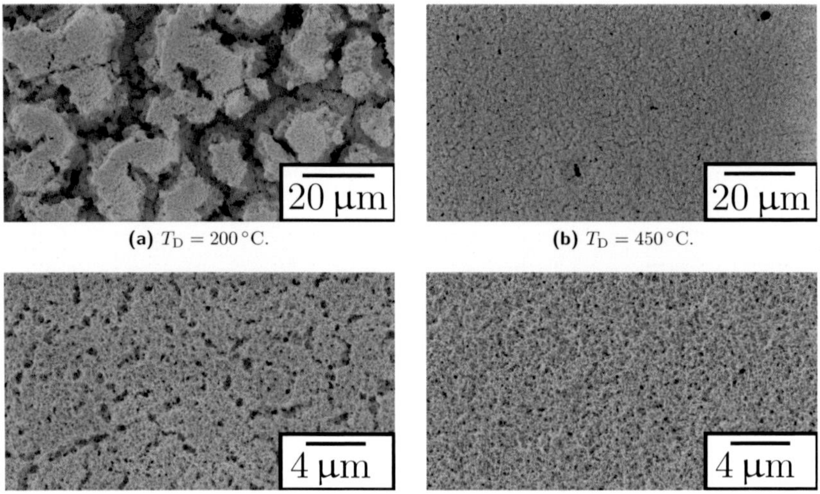

(a) $T_D = 200\,^\circ C$. **(b)** $T_D = 450\,^\circ C$.

(c) $T_D = 500\,^\circ C$. **(d)** $T_D = 550\,^\circ C$.

Figure 3.2: SEM images of samples prepared at different drying temperatures T_D [47].

Further investigations show that the drying temperature has a major influence on the film surface. In [46], it was found that the heating rate also influences the formation of cracks. To investigate this, samples with T_D from 200 °C to 650 °C are produced. T_D is increased in increments of 50 °C.

Figure 3.2 shows SEM pictures of exemplary film surfaces produced with ten material layers, each made using a rotation speed of 30 rps during spin coating for 20 s. Each subsequently applied layer is dried for five minutes at the given temperature T_D. Figure 3.2(a) shows cracks in the surface of a film dried at 200 °C. Furthermore, the material has a low adhesion to the substrate and peels off easily. Due to the cracks, the produced film is inhomogeneous. It is therefore unsuitable for the design of microwave devices. The sample depicted in Figure 3.2(b) is prepared using a drying temperature of 450 °C for each layer. The film shows large pores but the formation of cracks is reduced considerably in comparison to the sample dried at 200 °C. The size of the pores can be reduced, as seen in Figure 3.2(c), by using a drying temperature of $T_D = 500\,^\circ C$. The sample prepared with a drying temperature of 550 °C is depicted in Figure 3.2(d). It has a homogeneous surface, a good

(a) 30 rps. **(b)** 50 rps.

Figure 3.3: Samples prepared with 70 subsequently deposited layers at different rotation speeds.

adhesion to the substrate, and a comparably high film thickness. No further improvements were achieved in these experiments by utilizing higher drying temperatures.

As will be discussed in Chapter 4, ferrite films with higher thickness are desirable for the design of nonreciprocal components. To produce them, it is necessary to have either more subsequently deposited layers or a higher thickness of the individual layers. The thickness of the layers can be influenced by several process parameters, such as the rotation speed, the viscosity of the solution, and the substrate size.

Figure 3.3 shows films consisting of 70 layers dried at 550 °C. The sample shown in Figure 3.3(a) is produced with a rotation speed of 30 rps. To avoid cracking of the thicker films starting at the 20th layer, the substrate was cooled slowly back to room temperature after the drying of each layer. The produced film with a thickness close to 20 µm shows cracks and various defects. The sample shown in Figure 3.3(b) consists of 70 layers deposited at 50 rps. The surface shows less cracks, but defects can still be seen and the overall film thickness is 10 µm, which is close to the thickness achieved for films consisting of 20 layers deposited at 30 rps. Those films show less defects in comparison to the films consisting of 70 layers and the production is less time-consuming. Therefore, the slower rotation speed of 30 rps is used to produce films with 10 µm thickness, a crack-free surface, and good adhesion to the substrate.

Since cracks form at defects, these need to be avoided in order to produce films with a thickness larger than 10 µm. Particles contaminating the solution and the film surface result in defects and thus, assumably, a clean room and a suitable oven could improve the obtained results. Since no clean room was available for this work and the used high temperature laboratory oven has ceramic heating plates, which are not suitable for a clean environment, the production of films with a thickness higher than 10 µm is not further investigated in this work.

3.2 Elimination of Secondary Phases

To determine whether the produced films are single-phase BaM, the formation of secondary phases in the material is analyzed by XRD. For this, different samples were prepared and the process parameters were varied. If not mentioned otherwise, a drying temperature of $T_D = 550\,°C$ and a rotation speed of 30 rps is used in the following.

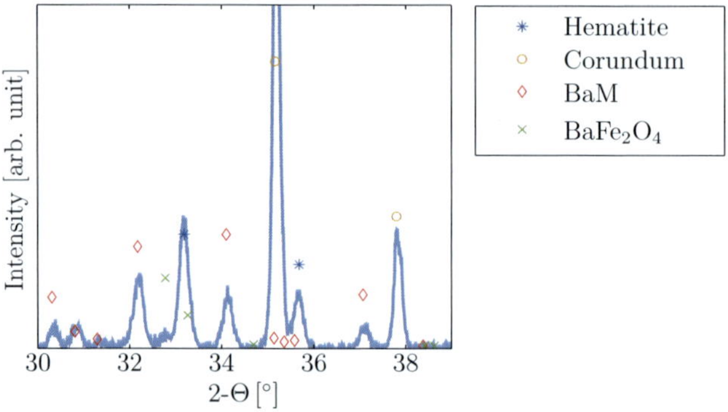

Figure 3.4: XRD-analysis of sample prepared with precursor ratio of 12.

Figure 3.4 shows the result of an analysis performed on a sample sintered at 850 °C. The weight of the precursor materials in the solution is chosen to obtain the stoichiometric Fe/Ba ratio of 12. The observed diffraction peaks can be matched to reference peaks for barium hexaferrite, hematite (Fe_2O_3), and corundum (Al_2O_3) powder samples. The peaks matched to corundum stem from the underlying substrate material. The hematite phase in the material indicates excess iron in the produced film. Nonmagnetic hematite in the film reduces the saturation and remanent magnetization of the material, which are crucial for the application in self-biased nonreciprocal components. Therefore, the formation of secondary phases in the film needs to be avoided. The chosen Fe/Ba ratio of 12 for the formation of BaM ($BaFe_{12}O_{19}$) cannot explain the observed iron excess in the film. To investigate this problem, an EDXS-analysis of the sample cross section is performed. For this, DuPont 951 LTCC tape was used as substrate material. To enable an analysis of the cross section, the sample was broken along slits after sintering the applied film.

Figure 3.5 shows the results of the performed EDXS-mapping. In Figure 3.5(a), an SEM image of the examined sample cross section is depicted. It can be seen that parts of the picture are out of focus. This is due to the roughness of the breaking edge. Since the sample surface needs to be at a certain working distance from the X-ray detector, it also influences the performed EDXS-analysis.

The map sum spectrum of the sample shows characteristic peaks that are matched to the detected elements, which in this case are C, Ca, O, Fe, Ba, Al, Si, Au, Pb, Cl, Ag, and K. As described in Section 2.4, as part of the sample preparation, a thin gold film was sputter-deposited onto the breaking edge to increase the conductivity of the sample surface. This is the reason for the detected gold.

Figure 3.5(b) shows a map of the detected iron in the material corresponding to the analyzed sample section. It can be seen from this that some iron is detected in all sample regions, but it is concentrated, as expected, in the region of the film on top of the substrate. Due to diffusion, there is also some iron detected in the substrate and few detections are seen in the region that is out of focus above the sample.

The map of the detected barium is depicted in Figure 3.5(c). It shows that the diffusion of barium into the substrate is stronger than that of iron. This strong diffusion is a possible reason for the iron excess in the film, which leads to the formation of hematite in the produced material. To avoid this iron excess, and thus prevent the formation of hematite, the ratio of iron to barium in the precursor solution is adjusted in the following.

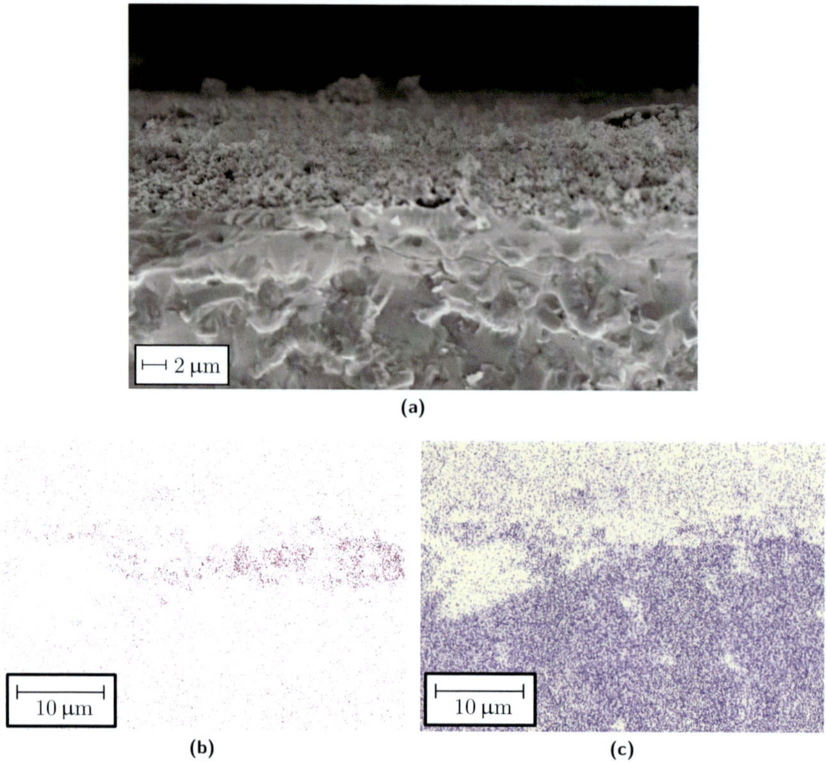

Figure 3.5: Results of an EDXS-analysis performed on the sample cross section depicted in (a). Picture (b) shows a map of the detected iron and (c) a map of the detected barium.

3.2.1 Stoichiometric Ratio in Precursor

To achieve the stoichiometric ratio of barium hexaferrite ($BaFe_{12}O_{19}$) in the film, the ratio of iron to barium in the solution is adjusted. This ratio (Fe/Ba) is called r in the following. The formation of secondary phases depends on almost all process parameters, amongst them the substrate material and the film thickness. In literature, values between $r = 2.28$ and $r = 12$ are reported to be suitable for the preparation of single-phase barium hexaferrite films on different substrate materials [32], [36]. In the context of this work, XRD-analyses are used to examine the formation of secondary phases in samples with r from 3 to 12.

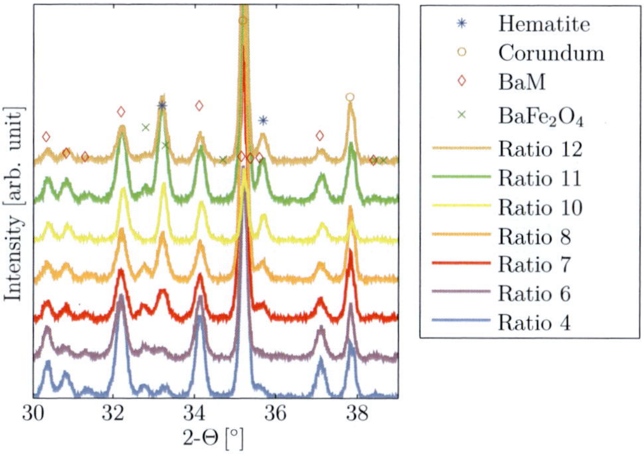

Figure 3.6: XRD-analysis results of samples with different r in the precursor solution sintered at $T_S = 850\,°C$.

Figure 3.6 shows the results of XRD-analyses performed on samples with $4 \leq r \leq 12$. As is commonly done, the intensity result is given in arbitrary units. This is reasonable, since it depends on the exact setup and equipment used in the experiment. The background is subtracted from the obtained result. For ease of comparison an intensity offset is added in the figures presenting the results of multiple analyses, to shift them against each other and make them easily distinguishable. The samples are sintered at 850 °C for 3 hours, 20 layers are produced, and each layer is dried at 550 °C. BaM and corundum are present in all examined films, whereas hematite (Fe_2O_3) and barium monoferrite ($BaFe_2O_4$) are detected in some samples.

The results show that none of the prepared samples is entirely free of secondary phases. The sample with $r = 12$ shows peaks for hematite, which decrease for samples prepared with lower values of r. The specimen with $r \leq 8$ exhibit peaks for barium monoferrite as well. Hematite is not detected in samples with $r \leq 6$. This indicates a barium surplus in the films with low ratios in the precursor solution. In an attempt to eliminate these secondary phases, six samples with $5 \leq r \leq 7$ are produced with similar process parameters except for a higher sintering temperature of 1050 °C.

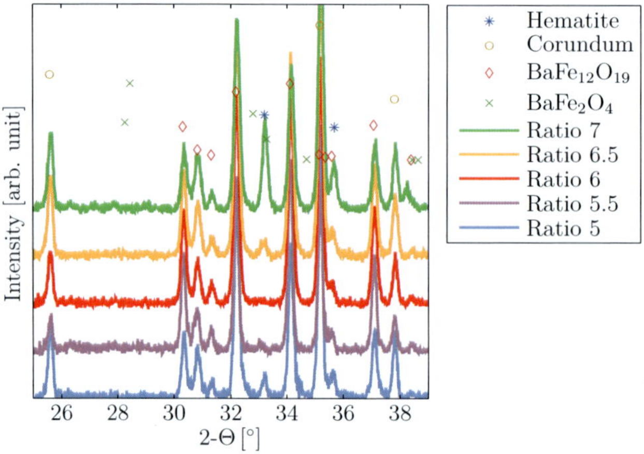

Figure 3.7: XRD-anlysis results of samples with different r in the precursor solution sintered at $T_S = 1050\,°C$ [47].

The results are shown in Figure 3.7. It can be seen that the sample with a ratio of $r = 7$ shows maxima related to hematite. The observed peaks decrease with lower r and no hematite is detected in the sample prepared with $r = 6$. Utilizing values of $r < 6$ again led to the formation of secondary phases in the film. Therefore, it can be concluded that for this method with the given process parameters, the optimum ratio of iron to barium in the precursor solution is 6.

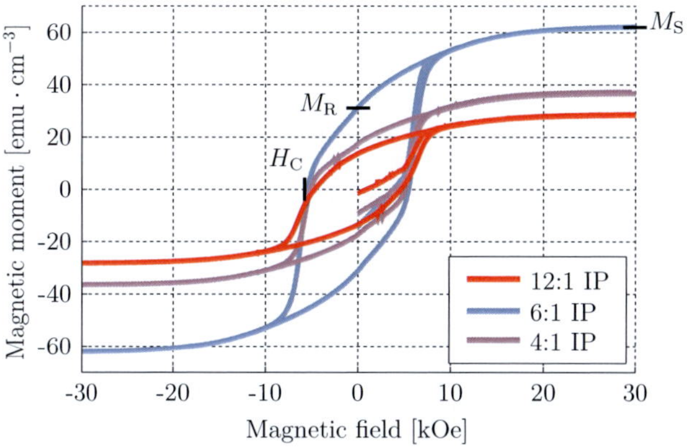

Figure 3.8: VSM-analysis results of samples with different r in the precursor solution.

To study the magnetic properties of the films prepared with different r, VSM-analyses are performed as described in Section 2.4. Figure 3.8 shows the results of IP measurements performed on films with the ratios of 12, 6, and 4. The precursor ratio significantly influences the saturation magnetization of the produced film. This matches well with the previous observations of the formation of nonmagnetic secondary phases depending on the used r. The measured saturation magnetization of the film with $r = 6$ is 62.1 emu·cm^{-3}. The values of saturation magnetization reported in literature for BaM films cover a large range, but are in general lower than the literature value for bulk material of 380 emu·cm^{-3}. A value of 80 emu·cm^{-3} is reported in [48] and 235 emu·cm^{-3} is the saturation magnetization obtained for the best film in [30]. In [38], a value of 122.6 emu·cm^{-3} was measured, and in [35], a saturation magnetization of 185.2 emu·cm^{-3} is reported.

In comparison, the films produced in this work exhibit a rather low saturation magnetization. Possible reasons for this are low densities of the produced films, diffusion of Al from the substrate into the film, which is known to decrease the saturation magnetization [49], or an overestimation of the film thickness due to the roughness of the film surface.

For the film prepared with $r = 4$, the saturation magnetization is reduced to 60 % of the value for the film with $r = 6$. If the stoichiometric ratio of $r = 12$ is used in the precursor solution, the saturation magnetization is reduced further to only 46 % of the result for a film prepared with $r = 6$. This matches well with the significant influence of the Fe/Ba ratio on the saturation magnetization of the films reported in [32], [36].

The squareness of the hysteresis loop ($M_\mathrm{R} \cdot {M_\mathrm{S}}^{-1}$) is only weakly influenced by r and is reduced from 0.50 for the film with $r = 6$ to 0.47 with $r = 4$. Since the films are intended for utilization in the remanent magnetization state, a higher saturation magnetization and consequently a larger remanent magnetization are favorable. This issue will be discussed further in Section 3.5.

The coercivity H_C of the hysteresis loop also decreases for the films containing secondary phases from 5.44 kOe to 5.09 kOe and 4.61 kOe for $r = 6$, $r = 4$, and $r = 12$, respectively. In [34], a value for the IP coercivity of 5.18 kOe was reported for a film with a Ba/Fe ratio of 0.1 and a sintering temperature of 850 °C. The value reported in [35] is 5.2 kOe for nanocrystalline particles annealed at 700 °C with a Fe/Ba ratio of 8. These results match well with the coercivities measured in this work. In the following, a ratio of $r = 6$ is used, if not stated differently.

3.2.2 Sintering Temperature and Duration

The sintering temperature and duration have a significant influence on the formation of secondary phases in the produced material. In order to integrate barium hexaferrite film into LTCC substrates, a sintering temperature below 900 °C is required. At higher temperatures, the LTCC is affected and changes its dielectric properties permanently [29]. To investigate the onset temperature for the formation of BaM while sintering the material, a DSC-analysis in combination with a TGA-analysis is performed.

For this, an alumina crucible is repeatedly filled with a precursor solution which is subsequently dried at 200 °C. This low temperature is used since the purpose is not to produce a crack-free thin film, but to fill the crucible. Drying the solution at 550 °C prior to the analysis would distort the result up to this temperature. However, drying at a sufficiently high temperature of 200 °C is necessary to fit more material into the crucible. Filling it

Figure 3.9: Filled crucible for DSC and TGA.

once and performing the experiment starting at room temperature yields a sample mass that is too small to perform the analysis. The sample prepared in this way is depicted in 3.9. Because the sample is heated prior to the experiment, it will only yield meaningful results from 200 °C to 900 °C which is the highest temperature that is achievable with the available equipment. The sample is heated from room temperature to 900 °C with $0.5\,°C \cdot min.^{-1}$ in synthetic air (20 % O_2, 80 % N_2).

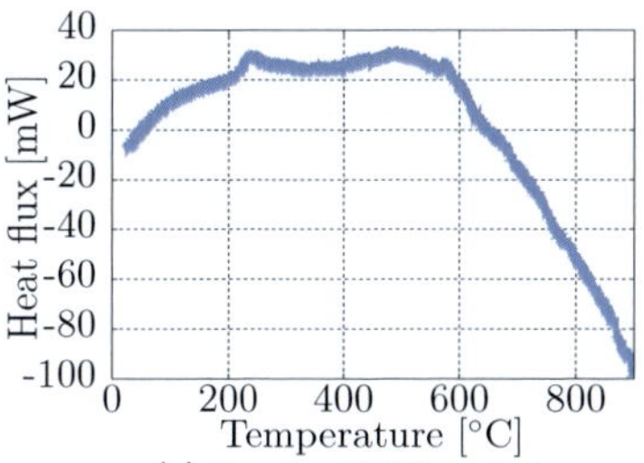

(a) Result of DSC-analysis.

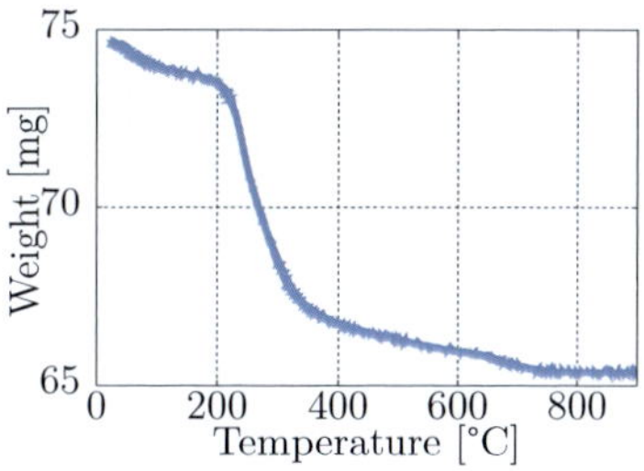

(b) Result of TGA-analysis.

Figure 3.10: DSC-analysis and TGA-analysis.

Figure 3.10 shows the outcome of the performed analysis. The result of the DSC in Figure 3.10(a) shows an exothermic reaction starting at 600 °C. It is reasonable to assume that this reaction is the onset of the BaM crystallization. This matches with the poorly crystallized particles reported in [50] with a sintering temperature of 600 °C. The reaction is not completed by the time the maximum temperature of 900 °C is reached. The sample was isotherm at 900 °C for 30 minutes after the heating. In this time, no significant change in heat flux was observed. From this, it can be concluded that the formation is not finished in 30 minutes at 900 °C. This result leads to the conclusion that longer sintering durations or higher temperatures are necessary, but it also shows that the formation of BaM begins at 600 °C, which is a suitable temperature for integration into LTCC.

The TGA in Figure 3.10(b) shows a significant reduction of weight at temperatures below 400 °C. One has to keep in mind that the small reduction of weight below 200 °C is due to drying the solution in the crucible prior to the experiment. Above 200 °C, the sample exhibits significant weight loss. This result matches well with the high drying temperature required for every layer to produce crack-free films as described in Section 3.1. The strain in the film is correlated with weight loss during drying. Therefore, drying every thin layer at 550 °C avoids strain in the prepared film.

To examine the influence of the sintering temperature on the formation of secondary phases, samples are prepared and sintered for 3 h at temperatures from 650 °C to 1050 °C. The samples are heated from the drying temperature of the last layer to the sintering temperature with $10\,°C \cdot min.^{-1}$.

Results of XRD-analysis performed on these samples are depicted in Figure 3.11. The sample sintered at 650 °C shows maxima for barium monoferrite and hematite. Only the main peaks of BaM can be identified. These results match well the result of the DSC-analysis which indicated that the formation of BaM begins at temperatures below 650 °C. For the sample sintered at 700 °C, the BaM peaks can be easily identified, but secondary phases are still present. For sintering temperatures of 750 °C and higher, no hematite, but only barium monoferrite is detected as a secondary phase. A reason for this is the chosen precursor ratio of $r = 5$, which contains slightly too much barium, but is therefore suitable to observe the influence of the sintering temperature on the formation of secondary phases. The samples sintered at 900 °C and 1050 °C show no secondary phases, but these temperatures are too high for the potential integration into LTCC substrates.

To examine the static magnetic properties of the films, VSM-analyses are performed. Three similar films are prepared and sintered for 3 h at 650 °C, 850 °C, and 1050 °C, respectively. The measured IP hysteresis loops of these films are depicted in Figure 3.12. As expected, the film sintered at 850 °C shows the previously observed hysteresis loop. The sample sintered at 650 °C shows a low saturation magnetization due to the contained nonmagnetic phases that were also detected in the XRD-analysis of the film on the alumina substrate.

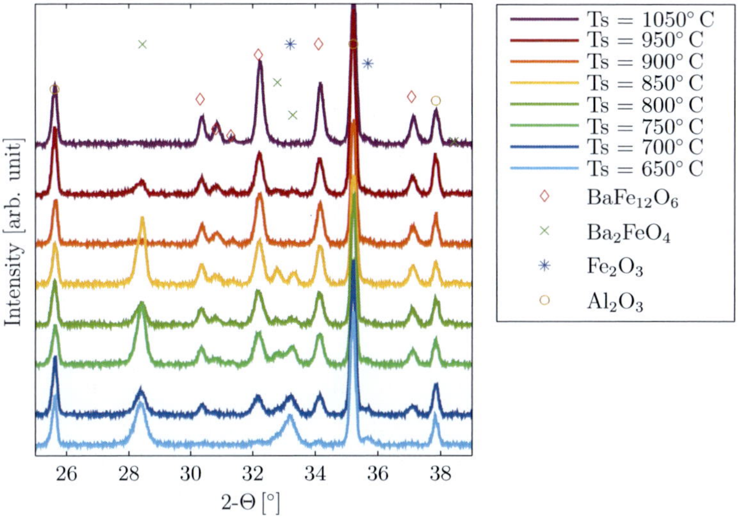

Figure 3.11: XRD-analysis results for samples with different sintering temperatures [47].

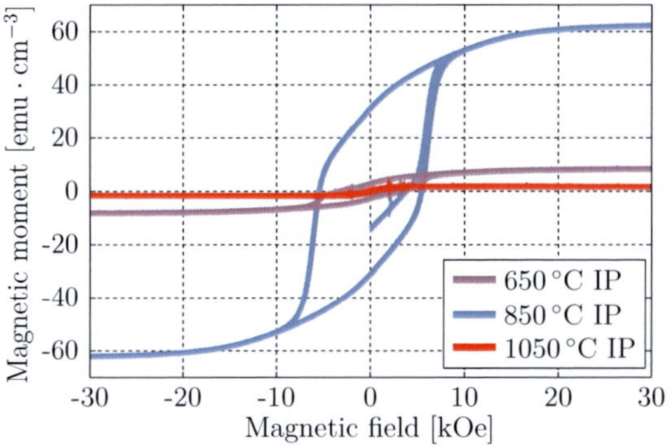

Figure 3.12: VSM-analysis results of films prepared on LTCC substrates sintered at different temperatures.

According to XRD-analyses performed on films prepared on corundum substrates, the material sintered at 1050 °C should contain no secondary phases. However, this temperature is not suitable for the LTCC substrate [29]. As can be seen from this result, the examined sample is nonmagnetic. Positioning the sample in the VSM utilizing a magnetic field is not possible, due to the low magnetization of the material. Therefore, the sample has to be positioned by hand. An XRD-analysis performed on this sample shows that the produced film is hematite. These results confirm that a sintering temperature of 1050 °C is advantageous for films prepared on corundum substrates, but that it is not suitable for films deposited on LTCC.

Next to the sintering temperature, the duration of sintering also has a significant influence on the formation of secondary phases in the film. To show this, a sample with $r = 5$ is prepared and sintered at 850 °C for 3 h. Based on the previous experiments, it is expected that this sample contains barium monoferrite and that, therefore, the influence of the sintering temperature on the barium monoferrite phase can be analyzed. A similar sample prepared from the same solution is sintered for 12 h.

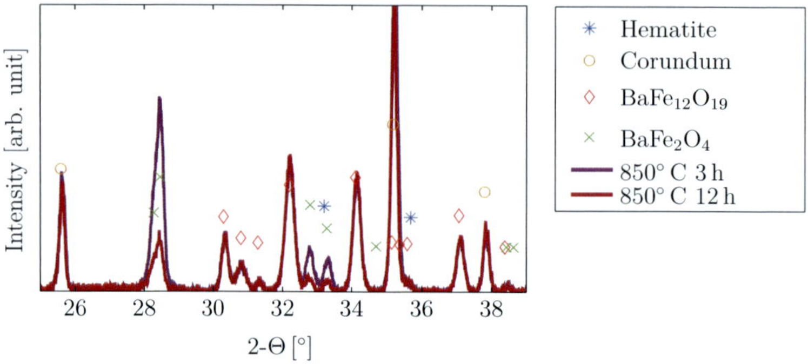

Figure 3.13: XRD-analysis results for samples with different sintering durations [47].

From the XRD-analysis results of these samples shown in Figure 3.13, it can be seen that both samples contain BaM and barium monoferrite. Since the prepared samples are similar, the peak heights can be used as an indicator for the quantity of secondary phases in the films. From this, it can be concluded that the film sintered for 12 h contains significantly less barium monoferrite than the film sintered for 3 h. One has to keep in mind that both films are prepared using the nonoptimal ratio of $r = 5$ in the precursor solution.

3.3 Measurements on Self-Biased Films

Microwave measurements without externally applied bias field are performed on the fabricated BaM material. The films are deposited directly on coplanar waveguides (CPWs) and the resulting S-parameters are measured. The CPW with a ferrite film deposited on top is sketched in Figure 3.14. The CPW is on alumina substrate with a thickness of 254 µm, the gold conductors have a thickness of 5 µm; the signal line marked with S in Figure 3.14 has a width of 250 µm, and the slot between signal and ground conductors marked with G has a width of 100 µm.

The magnetic field in the slot between signal and ground plane is elliptically polarized as indicated in Figure 3.14, with the sense of rotation depending on the direction of wave propagation on the CPW [51]. Therefore a nonreciprocal (NR) effect

$$\mathrm{NR} = |S_{12}|/|S_{21}| \tag{3.1}$$

is expected at the FMR frequency due to the resonance absorption.

Prior to the measurements, the ferrite film is magnetized to saturation. The direction of the applied field is in the substrate plane and perpendicular to the transmission line. The

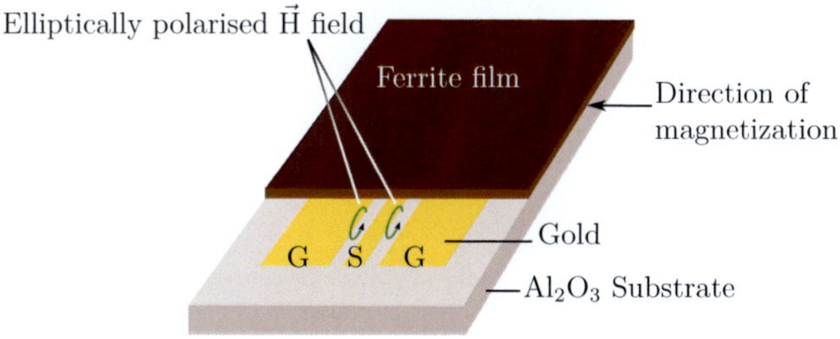

Figure 3.14: Sketch of CPW on an alumina substrate with ferrite film and elliptically polarized magnetic field between ground and signal conductors [47].

Figure 3.15: S-parameter measurements of a film prepared on a CPW.

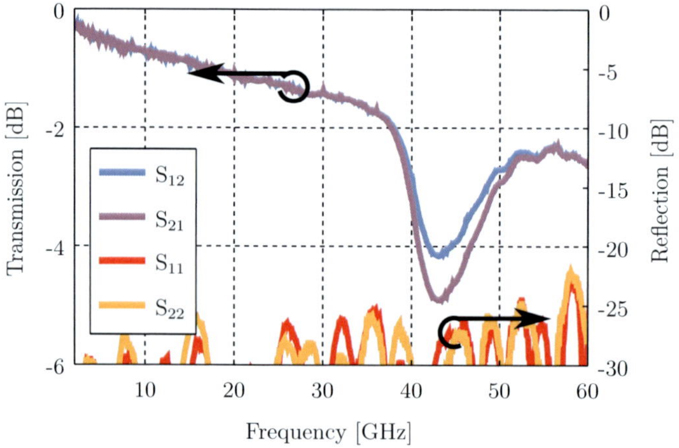

Figure 3.16: S-parameters of CPW with undoped barium hexaferrite film.

S-parameter measurements are performed in self-bias without externally applied magnetic field using I67 measurement probes from Cascade.

Figure 3.15 shows a film deposited on a CPW and the measurement probes. To enable contacting the CPW with probes, the ends of the gold conductors are etched as clean as possible without damaging the sample using heated hydrochloric acid.

Prior to the measurements, a calibration with the reference planes at the edges of the ferrite films is performed. The S-parameter measurement results of the BaM film on a CPW are plotted in Figure 3.16. The losses induced by the FMR can be seen around 43.2 GHz without any externally applied biasing field. Also, an NR effect of 0.76 dB can be seen. According to [13], the anisotropy field H_A of bulk BaM is 17 kOe, thus, using (2.4) the resonant frequency is expected to be 47.6 GHz. Thus, the FMRF measured for the film in this work is slightly lower than would be expected for bulk material. However, this result shows that BaM films capable of self-biasing can be successfully produced by means of the sol-gel method, to implement nonreciprocal components.

3.4 Doped Barium Hexaferrite Films

Conventional nonreciprocal microwave components utilize soft magnetic ferrites and permanent magnets to produce a static magnetic biasing field. As discussed in Section 2, soft magnetic ferrites have several disadvantages, but using a biasing field which is adjustable for a particular component instead of the anisotropy field of the material offers flexibility: The ferromagnetic resonant frequency can be adjusted by varying the applied field. In [2, p. 268], it is stated that the first step of testing a circulator is finding the operation point by tuning the biasing field. This is not possible if hard magnetic ferrites without externally applied bias field are used.

To change the FMR frequency of the hard magnetic ferrite material without applying an external biasing field, the anisotropy of the material needs to be adjusted. This is achieved by doping the material [52]. The produced hexaferrites can be modified by substituting some Fe^{3+} ions for other trivalent metal ions such as Al, Ga, In, Sc, As, Cr, or combinations of a tetravalent ion such as Ti^{4+}, Ru^{4+}, Zr^{4+}, or Sn^{4+} with a divalent ion such as Mn^{2+}, Ni^{2+}, Mg^{2+}, Co^{2+}, Zn^{2+}, Ti^{2+}, or Sn^{2+} [53]. The saturation magnetization, coercivity, anisotropy field, and FMR frequency of the barium hexaferrite are influenced by these substitutions [54].

In this work, the Fe^{3+} ions are substituted with scandium ions and aluminum ions in order to influence the FMRF. This is done in an attempt to manufacture suitable materials for nonreciprocal components with operating frequencies in the Ka-, Q-, and V-band.

3.4.1 Scandium and Aluminum Substituted Films

To produce barium hexaferrite films with lower anisotropy and therefore a lower FMRF, the iron ions are substituted with scandium. The scandium ions have a strong preference for the trigonal bipyramidal sites in the *R* block and the tetrahedral sites depicted in Figure 2.2 (in Section 2.1). The scandium ions begin to occupy the octahedral sites only for strong doping [55]. Scandium-doped barium hexaferrites prepared by conventional ceramic processing techniques are presented in [55], [56], and by PLD in [57] as well as in [58].

A higher FMRF can be achieved by substituting iron ions in the material with aluminum ions [59], [60]. This is done in [49] by using the so-called solution combustion technique to produce barium hexaferrite nanoparticles. Material analysis results for these particles are presented, but the FMRF is not measured. In [61], aluminum-doped barium hexaferrite single crystals are grown at a temperature of 1260 °C from carbonate flux and material analysis results are presented. In [62], the fabrication of aluminum-substituted barium hexaferrite films by means of the floating zone melting method is presented and measurement results for the FMRF of these films are provided. When doping with aluminum ions, the highest occupancy is observed on the octahedrally-coordinated sites and on the trigonal bipyramidal sites, while no aluminum ions are observed on the tetrahedral sites in the *S* block [61].

As discussed in Section 3.2.1, the optimum ratio of iron to barium in the precursor solution for the method used in this work is not the stoichiometric ratio of 12 but 6. To avoid the formation of secondary phases, the ratio of 6 is also used for the preparation of the scandium- and aluminum-doped barium hexaferrite films. Part of the iron is substituted and the amount of doping is characterized by the doping factors, which are in the following referred to as x and y for scandium- and aluminum-doping, respectively. The materials are prepared according to the sum formula $BaSc_xAl_yFe_{6-x-y}O_{19}$ instead of $BaSc_xAl_yFe_{12-x-y}O_{19}$. Therefore, the results for specific doping factors are not directly comparable to values reported in literature. The material is not doped with aluminum and scandium simultaneously, so either x or y is zero in each sample.

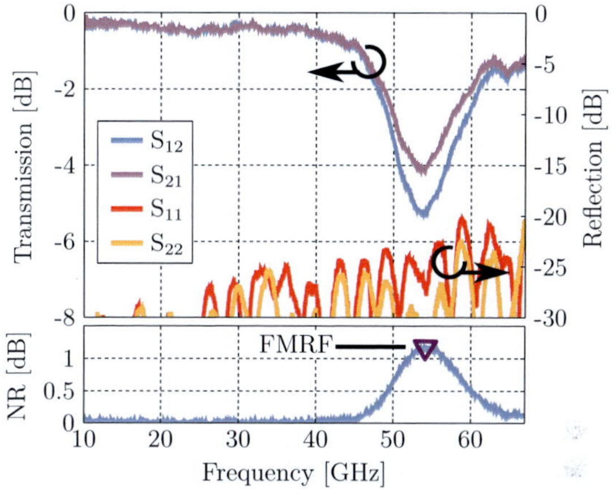

Figure 3.17: Measured S-parameters and NR effect of an aluminum-doped film ($y = 0.5$) [60].

To investigate the influence of doping on the FMRF, 23 films with doping factors from $x = y = 0.0$ to $x = 0.4$ and $y = 1$ are deposited on CPWs. All samples consist of 20 subsequently applied layers and are sintered for 3 h at 850 °C. All measurements are performed without an externally applied bias field.

Figure 3.17 shows the S-parameters, NR effect, and FMRF of an aluminum-doped sample with $y = 0.5$. From these results, it can be seen that the FMRF shifts from 43.2 GHz for the undoped sample to 54.4 GHz for the aluminum-doped sample with $y = 0.5$. For all prepared samples, similar measurements are performed and the FMRF and maximum NR effect are evaluated.

The measured FMR frequencies for the scandium- and aluminum-substituted films are depicted in Figure 3.18. These results show that the measured FMRF can be varied almost linearly within the two regions above and below the resonance of the undoped sample. The FMRF decreases almost linearly with increasing the scandium-doping factor from 43.2 GHz for the undoped sample to 27.1 GHz for the sample doped with $x = 0.4$.

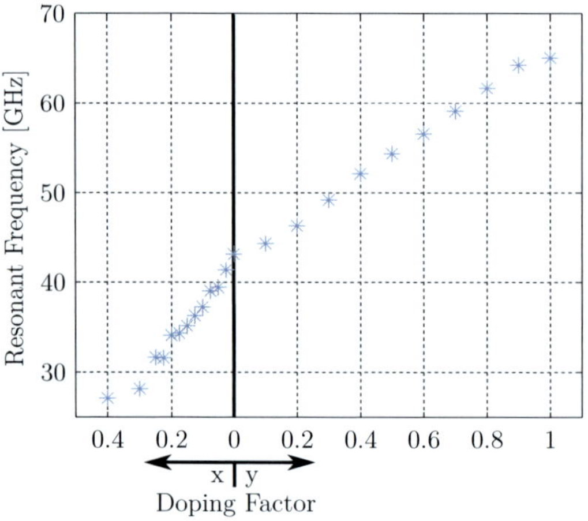

Figure 3.18: Measured FMRF of scandium- and aluminum-doped films with different doping factors [60].

The aluminum-doped sample with the highest doping factor of $y = 1$ shows the FMR at 65.1 GHz.

The influence of doping on the ferrite linewidth ΔH is shown in Figure 3.19. The permeability tensor is calculated from the measured S-parameters using the method presented in [63]. The ratio of the magnetic field components is determined from ANSYS Electronics Desktop 2017 and ΔH is obtained by fitting the extracted permeability tensor components using (2.21) and (2.22) with a complex resonant frequency

$$\omega_0 \rightarrow \omega_0 + j\alpha\omega \tag{3.2}$$

to account for losses. The damping factor α is related to the linewidth by

$$\Delta H = \frac{2\alpha\omega}{\mu_0\gamma}. \tag{3.3}$$

This yields only an indication of the linewidth, since the model is not suited for ferrites not magnetized to saturation. The obtained ΔH-values are normalized to that of the undoped

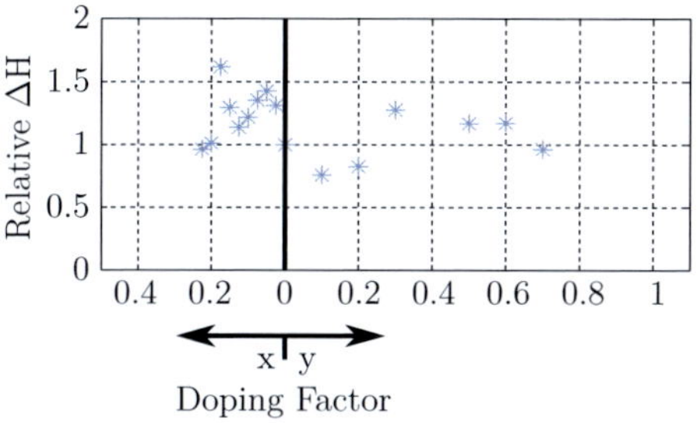

Figure 3.19: The influence of doping on the ferrite linewidth ΔH [60].

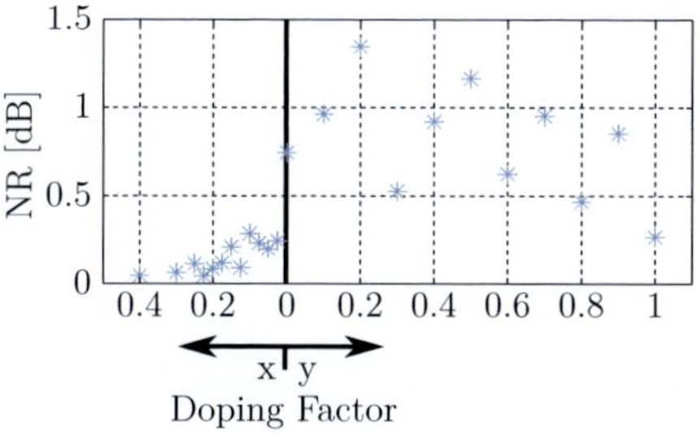

Figure 3.20: Peak NR effect of scandium- and aluminum-doped films with different doping factors [60].

film and presented here to indicate the influence of doping on the loss. For the samples with high doping factor the resonance is not completely within the measured frequency range and therefore, no values could be calculated by fitting. The results show no significant influence of doping on the ΔH-values. The variations are very likely due to measurement and manufacturing uncertainties with this simple setup [60].

The influence of scandium- and aluminum-doping on the measured NR effect is shown in Figure 3.20. The result for the peak NR effect varies strongly between samples. This is most likely due to inaccuracies in the etching of the ferrite film with hydrochloric acid. Deviations in length and thickness of the film influence the measured nonreciprocity significantly.

Table 3.1: Resonant frequencies (FMRF) and peak nonreciprocal effect (NR) of the differently doped ferrite films.

Sample	doping x	doping y	FMRF	$\|NR\|$
1	0.40	0	27.1 GHz	0.05
2	0.30	0	28.2 GHz	0.07
3	0.25	0	31.7 GHz	0.12
4	0.225	0	31.6 GHz	0.04
5	0.20	0	34.1 GHz	0.09
6	0.175	0	34.3 GHz	0.12
7	0.15	0	35.2 GHz	0.22
8	0.125	0	36.3 GHz	0.1
9	0.10	0	37.3 GHz	0.29
10	0.075	0	39.1 GHz	0.24
11	0.05	0	39.5 GHz	0.20
12	0.025	0	41.4 GHz	0.25
13	0	0	43.2 GHz	0.76
14	0	0.1	44.4 GHz	0.96
15	0	0.2	46.3 GHz	1.34
16	0	0.3	49.2 GHz	0.53
17	0	0.4	52.1 GHz	0.92
18	0	0.5	54.4 GHz	1.16
19	0	0.6	56.6 GHz	0.62
20	0	0.7	59.1 GHz	0.95
21	0	0.8	61.7 GHz	0.47
22	0	0.9	64.3 GHz	0.85
23	0	1.0	65.1 GHz	0.27

Nonetheless, the measurement results show that the NR effect decreases for the films doped with scandium. The sample with doping factor $x = 0.4$ shows an NR effect of 0.05 dB, which is close to the measurement noise. Because of this small nonreciprocity no higher doping factors, and therefore lower FMRF, are investigated. The effect is less pronounced for the aluminum-doped samples. Here, samples with a doping factor up to $y = 1$ are prepared and measured. The measurements are performed from 10 GHz to 67 GHz. Therefore, samples with a higher doping factor are not produced since the FMRF is expected to be close to or above 67 GHz. The results indicate that higher doping factors are possible. Table 3.1 summarizes the measured FMRF and peak NR results for all doped samples and one undoped sample.

3.4.2 Material Analysis of Substituted Films

As already mentioned above, in addition to the anisotropy field and FMR frequency also the saturation magnetization and coercivity of the BaM films are influenced by these substitutions [54]. XRD-analyses and VSM-analyses of the doped material are performed to investigate these effects and the formation of secondary phases due to these substitutions.

X-Ray Diffraction Analysis

A possible reason for the reduction of the NR effect in the scandium-substituted films is the formation of secondary phases due to the doping. The results for an analysis of a film with

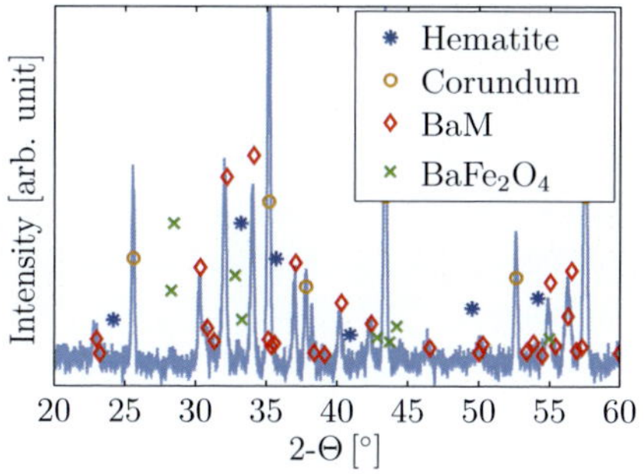

Figure 3.21: XRD-analysis result of scandium-substituted film with $x = 0.3$ [52].

the doping factor $x = 0.3$ is depicted exemplarily in Fig. 3.21. From this, it can be seen that the film with a rather low resonant frequency of 28.2 GHz does not contain secondary phases. It can therefore be concluded that the formation of nonmagnetic secondary phases is not the reason for the inferior performance of films with higher doping factors.

Similar XRD-measurements are performed on the aluminum-doped films and the results are depicted in Figure 3.22. The results for the films with $y = 0.3$, $y = 0.6$, and $y = 0.8$ show that barium monoferrite forms in all analyzed aluminum-doped samples. This problem could not be resolved by using higher sintering temperatures. Also, an adjustment of the ratio of barium to iron in the precursor solution from 6 to 7 and 8 does not eliminate the

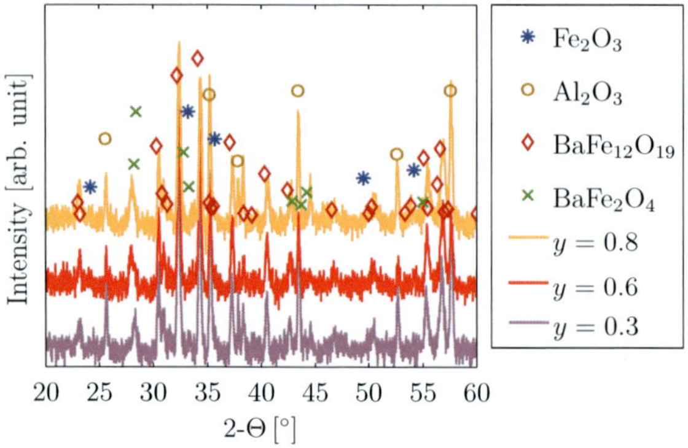

Figure 3.22: XRD-analysis results of aluminum-substituted films with different doping factors [60].

barium monoferrite. The sample prepared with a ratio of 7 contains barium monoferrite. The ratio of 8 leads to the formation of hematite as well as of barium monoferrite. The XRD-analysis results presented in [49] for aluminum-substituted barium hexaferrite also show single-phase barium hexaferrite for the undoped sample and a secondary phase for samples substituted with aluminum.

Vibrating Sample Magnetometer Measurements

When substituting nonmagnetic scandium or aluminum cations in the material for the Fe^{3+} with $5\,\mu_B$, the saturation magnetization of the material is expected to decrease as described in [55]. As stated before, the doping factors are not comparable to results reported in literature due to the differences in the manufacturing process of the material, but the effect is expected to be quantitatively the same. To investigate this for the films produced in this work, VSM-measurements are performed.

Figure 3.23 shows the analysis results for films doped with $x = 0.1$ and $y = 0.5$ in comparison to an undoped sample. From these results, it can be seen that the saturation magnetization of the scandium-doped material is reduced to 24 % of the value for the undoped film. Since the produced material is intended for use in the remanent magnetization

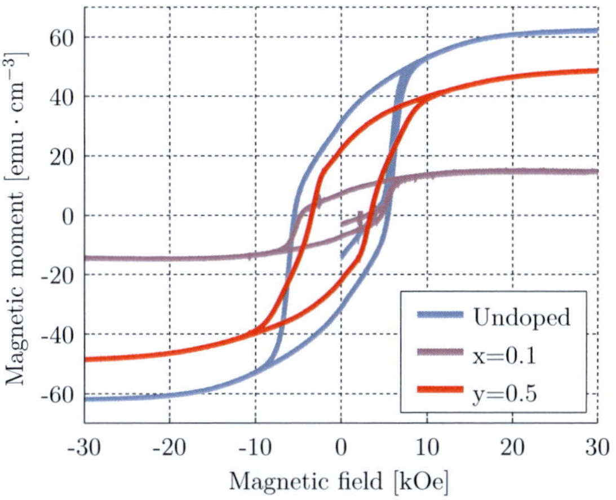

Figure 3.23: VSM-analysis results for samples doped with scandium $x = 0.1$ and aluminum $y = 0.5$ in comparison to an undoped sample [60].

state, the squareness ($M_R \cdot {M_S}^{-1}$) of the measured hysteresis loop is an important characteristic of the material. It changes only slightly from 0.5 for the undoped film to 0.48 for the doped film, indicating that the remanent magnetization is reduced along with the saturation magnetization in the scandium-doped films.

In comparison to the undoped material, the saturation magnetization of the aluminum-doped film is only reduced to 77 %. The squareness is 0.45 and therefore lower than that for the undoped and the scandium-doped films. One explanation for the reduction in squareness and saturation magnetization could be the barium monoferrite phase in the film. Nevertheless, the remanent magnetization of the aluminum-doped sample is significantly higher than that of the scandium-doped film.

The coercivity of the material is reduced from 5.5 kOe for the undoped film to 4.7 kOe and 3.4 kOe for the film substituted with scandium and aluminum, respectively. For the aluminum-doped barium hexaferrite nanoparticles presented in [49], the coercivity is reduced from 5.7 kOe for the undoped sample to 4.8 kOe for the sample with doping factor 1. The iron content in our precursor solution is reduced from 12 to 6 parts. Therefore, the doping factor of 1 from [49] may be most suitable for comparison to a doping factor of $y = 0.5$ in this work. Therefore, it can be concluded that the trend of reduced saturation magnetization and coercivity observed in this work matches the results reported in [49].

The reduction in remanent magnetization for the scandium-substituted material could be the major reason for the reduced nonreciprocal effect. A possible way to improve the results would be to produce a textured material, as will be discussed in Section 3.5. Also, an increased density of the produced films could improve the measurement results for both doped and undoped films.

3.5 Textured Barium Hexaferrite Films

Texture in BaM films, meaning a preferred c-axis orientation of the crystallites in the film, is important to achieve a high permanent magnetization of the material without an externally applied magnetic field. Thus, the squareness $M_{\mathrm{R}} \cdot {M_{\mathrm{S}}}^{-1}$ of the hysteresis loop is high in the direction in which the c-axis of the crystallites are aligned. This high remanent magnetization is a crucial material property for the performance of self-biased components. Texture of a film can be measured, for example, by VSM-measurements of the IP and OP hysteresis loop, by a peak height analysis in XRD-measurement, or a torque measurement in an externally applied magnetic field.

There are several processes by which textured BaM films can be produced. In [64], PLD is used to fabricate textured BaM films with a squareness of up to 0.95 on a-plane sapphire substrates. The orientation of these films depends on the conformity of the sapphire substrate orientation and the preferred orientation in the produced film. For films with thicknesses above 5 µm, fractures and delamination are observed. This issue is broached in [24], where a highly oriented seed layer is deposited by PLD on oriented sapphire substrates and a film with larger thickness is grown on this layer by LPE. The squareness of the hysteresis curve for these films decreases with increasing film thickness from 0.9 at 1 µm to 0.3 at 20 µm. A BaM film with 7 µm thickness and a squareness of 0.84 is presented as a good trade-off in that work. In [65], BaM films which also depend on the substrate material for their orientation are sputter-deposited on thermally-oxidized silicon substrates using a Pt interlayer with 50 nm thickness. The oxygen pressure during deposition is controlled to achieve a squareness of up to 0.95 for a film with 40 nm thickness decreasing to 0.65 for a film with a thickness of 120 nm.

Another method to produce textured BaM films is screen-printing a paste consisting of BaM particles dispersed in a binder and drying the wet-screen-printed film at low temperatures (typically 150 °C–250 °C) in a magnetic field to orient the particles [27]. This method presents the advantage that the preferred orientation of the particles in the film does not depend on the substrate material. With this promising approach, films with a thickness of up to 500 μm and a squareness of up to 0.96 are produced. The films show densities in the range of 85 – 90 % after annealing at 1300 °C.

In a recent study [66], textured films are also produced by means of the sol-gel method. The used substrate material is layered Pt/TiO_2/SiO_2/Si and the films have a thickness of 350 nm. There are several differences in the production process used in [66] compared to the one used in this work. A major difference is in the drying of the single layers, which is performed using temperatures of 150 °, 250 °C, and 400 °C for 1 min., 4 min., and 4 min., respectively. These drying steps are followed by an annealing step at 700 °C in O_2 atmosphere for each deposited layer until the overall film thickness of 350 nm is achieved. In a last step, the film is annealed at 1000 °C for 2 h, again in an oxygen atmosphere. The produced films show a crystallographic alignment degree of 0.93 for an undoped film and 0.9 for a scandium-doped film. The orientation is determined from XRD-analysis results using the Lotgering method and not from VSM-measurement results for the hysteresis loop.

In comparison to the process used in this work, the film thickness of 350 nm is rather small, the sintering temperature of 1000 °C is too high for LTCC substrates. Furthermore, the high temperature step in an oxygen atmosphere requires a more sophisticated sintering setup and the texture of the film may depend on the used substrate material.

In the frame of this work, no textured BaM films are produced. The production of c-axis-oriented films was impeded by the requirement that the final films should be produced on LTCC substrates. Therefore, no substrate with a preferred orientation could be used. Also, the desired high film thickness is for most production processes in contradiction to the goal of highly c-axis-oriented films. The particles in the screen-printed films are aligned by drying at low temperatures in an externally applied magnetic field. This approach is not applicable for films produced by means of the sol-gel method, since the wet films contain only the nonmagnetic precursor materials and the BaM crystallites are not formed until the sintering step.

In an experiment using a superconducting magnet capable of producing a 5 T field in combination with a tube furnace, the BaM films were crystallized at 850 °C under the influence of the external magnetic field. This experiment did not yield textured BaM films and, due to difficulties with controlling the temperature of the tube furnace, the results are not comparable to those of the reference samples sintered without the influence of the magnetic field and are therefore not presented in this work. The main reason that the films sintered in this experiment show a random c-axis orientation is presumably that the sintering temperature of 850 °C is far above the Néel temperature of BaM, at which the material becomes paramagnetic. In [67], this temperature is determined to be 426.9 °C in the case of thin films. In the same study, the effective anisotropy is determined for various temperatures using torque measurement results. An effective anisotropy of 0 was measured at the Néel temperature, implying that magnetic annealing at high temperatures is not effective to achieve a preferred c-axis orientation of the BaM crystallites.

Thus, even though textured films would be preferable, no textured films are produced in the context of this work. The main impediments are the requirement of using LTCC substrate material, the required high film thickness, and the characteristics of the sol-gel method. The produced films with randomly oriented *c*-axes of the crystallites are thus used in the following.

3.6 Barium Hexaferrite Films on LTCC Substrates

Since the aim of this work is to produce small and lightweight nonreciprocal components, it is important that the produced ferrite films can be integrated into LTCC. For this, four commercially available LTCC tapes are chosen and tested as a substrate material for the produced barium hexaferrite films. The tapes are the 951 and 9k7 by DuPont, the A6M by Ferro, and the CT 702 by Heraeus. The sintered tape samples were kindly provided by MSE and the Ernst-Abbe-Hochschule Jena. In the following, these tapes are briefly

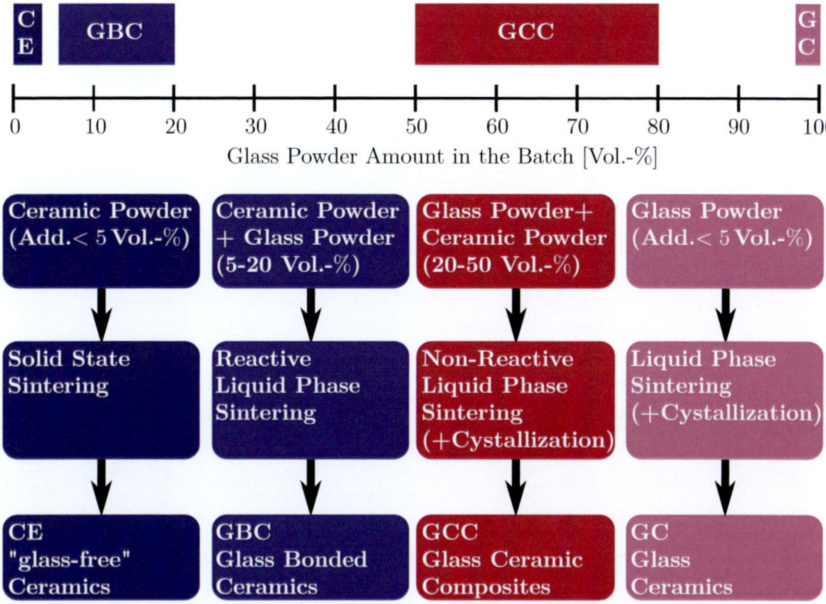

Figure 3.24: Overview of the composition, raw materials, and sintering mechanisms of the different LTCC types [68].

characterized and some differences which are important with regard to the production of ferrite films are highlighted.

Figure 3.24 shows the composition, raw materials, and the characteristic sintering mechanisms of four different LTCC types. The glass amount in most commercially available LTCC tapes is higher than 50 %. These tapes can be classified either as glass ceramic composites (GCCs) or glass ceramics (GCs), depending on the glass powder amount in the raw materials [68].

The Dupont tapes 9k7 and 951, as well as the Heraeus CT 702 tape are GCC tapes. All chosen GCC tapes contain 20 – 50 % Al_2O_3 filler and correspondingly 50 – 80 % glass matrix material. The main role of the glass phase is to serve as a bonding agent, to enable densification via viscous flow-dominated sintering [69]. During sintering, the glass densifies first, followed by devitrification to low-loss phases and forms a low-loss, dense glass-ceramic body [70]. This increases the mechanical strength of the substrate and the re-firing stability for post-firing processes in the temperature range of the sintering temperature, as needed for the processing of the ferrite films prepared in this work.

The Ferro A6M tape used in this work is a GC tape. It consists of a Ca-borosilicate glass without filler particles [71]. The glass enables densification and devitrifies almost completely during sintering. The properties of the resulting material are controlled by the degree of crystallization, which can be enhanced by adding a small amount of crystalline phase that acts as a nucleating agent [72].

Glass bonded ceramics (GBCs) are a relatively new class of LTCC. For this type, the low amount of glass (5 – 20 Vol.-%) serves as a low temperature sintering flux. The dielectric constant of these materials can be tailored very well due to the high amount of ceramic material and the variety of options for this ceramic filler [72]. The "glass-free" ceramic (CE) type of LTCC with a maximum of 5 Vol.-% additives uses solid-state sintering below the melting point of the ceramic powder and produces an almost glass-free material. It is used for the production of functional layers in multilayer LTCC [73]. In this work, only the named commercially available GCC and GC tapes are used. The production of a ferrite film on GBC or CE tapes is not attempted in this work.

Table 3.2: Peak sintering temperature, holding time at peak temperature, and microwave properties for the LTCC tapes used in this work as stated by the manufacturers.

Tape	DuPont 951	DuPont 9k7	Heraeus CT 702	Ferro A6M
LTCC type	GCC	GCC	GCC	GC
Sinter temp.	850 °C	850 °C	850 – 865 °C	850 °C
Hold at peak	20 min.	20 min.	20 – 30 min.	10 – 15 min.
Permittivity	7.8 ± 0.2	7.1 ± 0.2	7.5 – 7.9	5.9
Loss tangent	0.014 at 10 GHz	0.001 at 10 GHz	0.0021 at 1 kHz, 25 °C	< 0.002 1 – 100 GHz
Source	DuPont[a]		Heraeus[b]	Ferro[c]

[a] DuPont E.I. du Pont de Nemours and Company, DuPont GreenTape low temperature co-fired ceramic system, Design and Layout Guidelines, Nov. 2009.
[b] W. C. Heraeus GmbH & Co. KG, Low Temperature Cofireable Heratape.
[c] Ferro Electronic Materials, Low Temperature Co-fired Ceramic System, A6M High Frequency, High Reliability LTCC Tape System, Feb. 2013.

Table 3.2 lists properties of the LTCC tapes used in this work. The sintering temperatures for all tapes, as suggested by the manufacturers, are $\geq 850\,°C$ and are therefore higher or equal to the temperature required in a post-firing step to apply the barium hexaferrite film.

The permittivity and loss tangent as supplied by the manufacturers are also listed in Table 3.2. The LTCC materials are composite structures of glass and crystals and their dielectric properties depend on the constituent materials in the composite. The remaining glass phase after sintering lowers the relative permittivity and increases the dielectric loss [74]. It is therefore desirable to use a material that is almost completely devitrified and consists mostly of crystalline phases. Table 3.3 summarizes the results of XRD-analysis of the sintered tapes that are used in this work. The results are taken from [69], [75].

Table 3.3: Non-dispersive X-ray spectroscopy analysis results of sintered tapes [75], [69]. The major phases are listed first followed by the minor phases in italics.

Tape	DuPont 951	DuPont 9k7	Heraeus CT 702	Ferro A6M
Phase	$\alpha-Al_2O_3$	$\alpha-Al_2O_3$	$\alpha-Al_2O_3$	$CaSiO_3$
Structure	$Ca(Si,Al)_4O_8$	$LaBO_3$	$BaAl_2Si_2O_8$	
		$LaPO_4$		*$MgSiO_3$*
		$CaAl_2B_2O_7$	*SiO_2*	*SiO_2*
			$(Mg,Ca)SiO_3$	*CaB_2O_4*
		$La_7P_3O_{17}$		
		$AlPO_4$		

In traditional GCC tapes, as for example the DuPont 951, almost no additional self-derived or reaction-derived crystalline phases other than the original ceramic phase form during sintering. The major additive in the DuPont 951 is Al_2O_3. The amount of glass remains almost unchanged during sintering [69] and explains the rather high loss tangent of 0.014 at 10 GHz.

For the DuPont 9k7 tape the ceramic filler is Al_2O_3 as well. The unfired tape shows only peaks corresponding to the alumina filler. It can be seen from the results in Table 3.3 that the glass partially crystallizes to form self-derived crystalline phases (mainly lanthanum borate $LaBO_4$ and monazite $LaPO_4$) and forms reaction-derived crystalline phases (mainly calcium aluminum borate $CaAl_2B_2O_7$) with the alumina. The high amount of crystalline phases in the sintered material explains the rather low loss tangent of 0.001 at 10 GHz.

Figure 3.25: Photograph of DuPont 951 tape without (left) and with (right) barium hexaferrite film.

In the Heraeus CT 702, alumina is also used as a ceramic filler. During sintering, it forms celsian ($BaAl_2Si_2O_8$) as a reaction-derived phase. An EDXS-analysis in [75] shows that this tape contains five atomic percent of barium. This could influence or reverse the diffusion of barium from the film into the substrate. The loss tangent of 0.0021 at 1 kHz is provided by the manufacturer. Values for the permittivity at higher frequencies, of for example 6.8 at 10 GHz, are given in [76], but no value for losses at higher frequencies is given.

The Ferro A6M tape shows no crystalline phase prior to sintering, since it is a GC tape. It is a crystallizable Ca-borosilicate glass with no filler particles [71]. The major phase after sintering is wollastonite ($CaSiO_3$). The loss tangent in this tape is below 0.002 for frequencies below 100 GHz, and at 10 GHz the loss tangent is even below 0.001. It would therefore be desirable to use the Ferro A6M tape for the fabrication of low-loss microwave devices at high frequencies.

To produce BaM films on the chosen LTCC substrates, the process optimized on the alumina substrates is used. The ratio of iron to barium in the precursor solution is kept at 6, assuming a diffusion of barium into LTCC substrates similar to that for alumina substrate. The solution is homogenized and spin-coated as described in Section 2.3.2. The layers are dried for 5 minutes at 550 °C. 20 layers are applied subsequently and are then sintered for 3 h at 850 °C. The results for each tape are presented and discussed in the following.

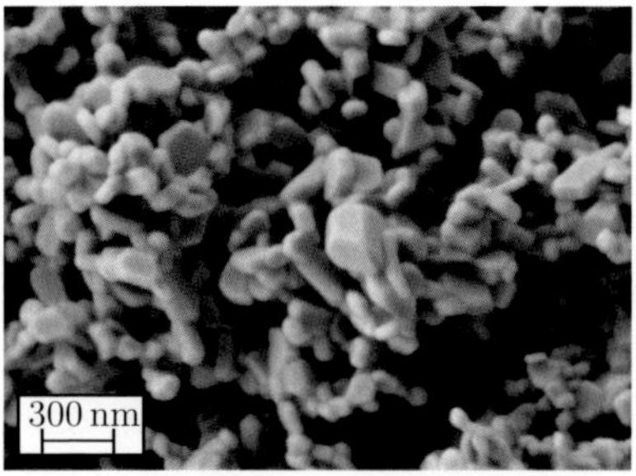

Figure 3.26: SEM image of produced ferrite film on DuPont 951 tape sintered for 3 h.

BaM film on DuPont 951

Figure 3.25 shows two DuPont 951 substrates, one with and one without barium hexaferrite film. The film shows no macroscopic cracks and good adhesion to the substrate.

To investigate the surface of the produced sample, an SEM image is taken and shown in Figure 3.26. The film surface is similar to the one shown in Section 3.1. It is crack-free except for the regions close to the substrate edges and corners. No large pores are observed on the surface, similar to the undoped film on alumina substrate dried at 550 °C shown in Figure 3.2(d). The crystal size as seen from the SEM images is on the order of 200 nm. As discussed in Section 2.4, measuring the density of the produced film is difficult due to the low weight of the film and uncertainties, which are caused by inhomogeneities in the film thickness. The film thickness varies especially at the substrate edges and the corners. Nevertheless, from the SEM image in Figure 3.26 it can be seen that the film is not sintered to full density. Usually, the density of the produced film increases with increasing sintering time or temperature [77]. Applying a sintering temperature higher than 850 °C is not an option while using LTCC substrates. Further devitrification of the remaining glass phase would take place and reactions between film and substrate could occur. In [29], it is also shown that DuPont 951 Tape changes its permittivity during a post-firing treatment at 900 °C. However, using a longer sintering time to produce a film with higher density is possible.

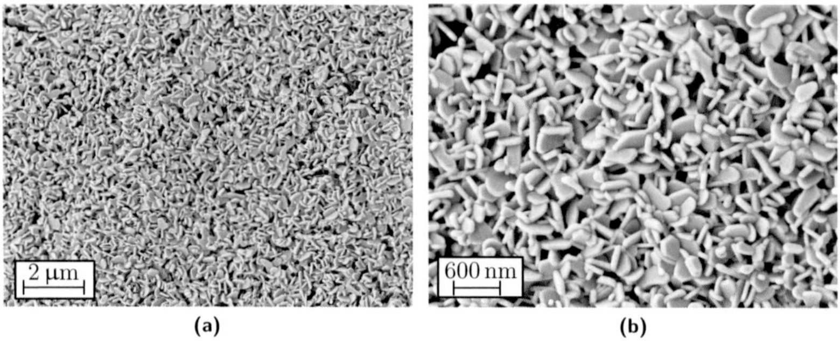

Figure 3.27: SEM images of a ferrite film produced on DuPont 951 tape with a sintering time of 15 h.

To investigate this effect, a film was prepared using the same process parameters except with a longer sintering time of 15 h. SEM images of the resulting film surface are depicted in Figure 3.27. In Figure 3.27(a), an overview of the film surface can be seen. It is crack-free and no large pores are observed. Therefore, it can be concluded that the longer sintering time did not result in an inferior surface quality.

Figure 3.27(b) shows a more detailed view of the film surface. It should be noted that the grain size did increase with longer sintering duration and is now close to 400 nm. The process of grain growth is considerably interconnected with the development and morphology of the pores in grain aggregates [77]. Indeed, even though the film density is not measured, it can be presumed from the SEM image in Figure 3.27(b) that the increased sintering time leads to a higher density in the produced film. This improved density is linked to the larger grain size.

To investigate the phase structure of the produced films, XRD-analyses are performed. The result for the film with a sintering time of 3 h is shown in Figure 3.28 for a large angular range. All peaks, which are expected for barium hexaferrite can be identified from the result. Barium monoferrite and hematite, which were often observed for films prepared under nonideal conditions before, cannot be detected. The XRD-analysis performed on the film with a sintering duration of 15 h shows a similar result. Two rather small peaks can be distinguished at angles of 25.6 ° and 28 °, but they do not match any of the expected

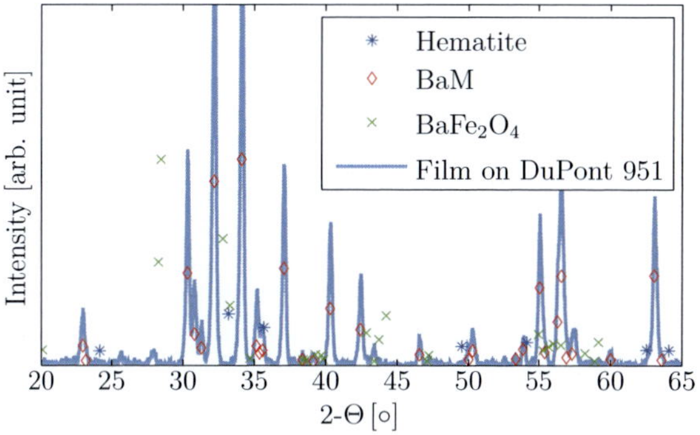

Figure 3.28: XRD-analysis result for barium ferrite film prepared on DuPont 951 LTCC tape and sintered for 3 h.

phases in the ferrite film. These peaks are confirmed to stem from the substrate by running a blank substrate, for which the same peaks are observed among other minor ones.

It can therefore be concluded that this method, with the used process parameters, is suitable to produce single-phase barium hexaferrite films on the DuPont 951 LTCC substrate. The fabricated film shows no secondary phases, no cracks, and has a good adhesion to the substrate.

BaM film on DuPont 9k7

To test the DuPont 9k7 tape as a substrate for the barium hexaferrite films, a film is produced using the same process parameters and again a sintering duration of 3 h. SEM images of the resulting film surface are depicted in Figure 3.29. An overview of the surface is shown in Figure 3.29(a). Small cracks can be observed all over the sample and one can be seen in the lower center region of Figure 3.29(a). Furthermore, small pores can be observed. In Figure 3.29(b), a more detailed SEM image of the substrate surface is shown. In this image, no barium hexaferrite crystals can be identified. The surface looks fundamentally different from the ones observed before. One possible reason for this are reactions of the substrate material with the prepared film during sintering. Also, a small

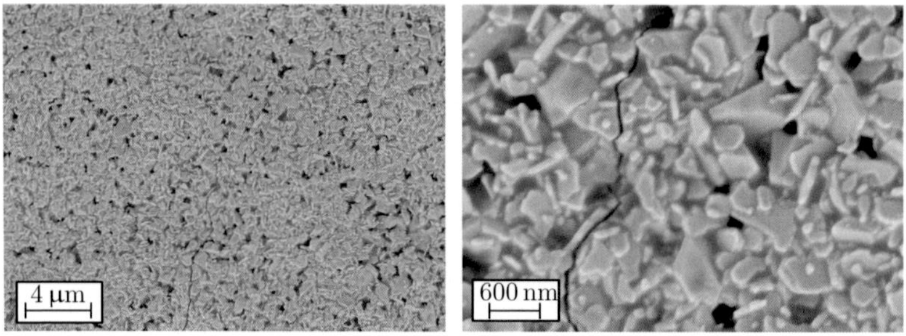

Figure 3.29: SEM images of a ferrite film produced on DuPont 9k7 tape.

crack as well as pores can again be seen in Figure 3.29(b). These major differences are only observed with the SEM. Macroscopically, the film surfaces of the samples depicted in Figure 3.27 and Figure 3.29 look similar. With a light microscope, some of the larger cracks of the film on the DuPont 9k7 substrate can be observed. From this result, it can already be concluded that the choice of DuPont 9k7 as a substrate material has a major impact on the produced film.

To investigate the phase structure of the film, an XRD-analysis is performed. From the result depicted in Figure 3.30, it can be seen that the expected peaks for barium hexaferrite, are detected. Next to BaM, hematite forms in the produced film, but no barium monoferrite is detected. Some additional peaks due to the substrate, for example corundum at 25.6 °, can also be observed in the intensity spectrum.

It can thus be concluded that with the used process parameters no usable barium hexaferrite film can be produced on the DuPont 9k7 substrate. One possible reason for this could be the remaining glass phase in the substrate, which may lead to liquid-phase sintering, explaining the fundamentally different surface structure observed in the SEM images. Also, as discussed above and shown in Table 3.3, unlike the DuPont 951 tape, the DuPont 9k7 forms self-derived as well as reaction-derived crystalline phases during the first sintering step. It is reasonable to assume that this tape also shows a higher reactivity to the applied film during a second sintering step which is performed at the same temperature. Still, barium hexaferrite formed on the sample surface, along with hematite. This could be due

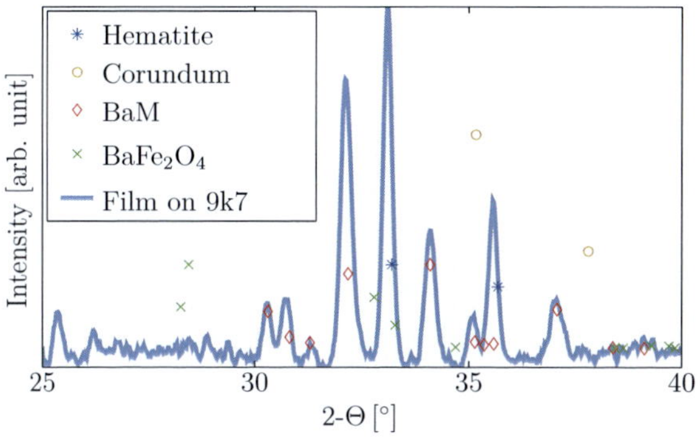

Figure 3.30: XRD-analysis result for ferrite film on DuPont 9k7 LTCC tape.

to a stronger diffusion of the barium into the substrate, leading to a lack of barium at the sample surface.

No further process optimization was attempted to produce a BaM film on this substrate. One promising option would be to use a lower sintering temperature in an attempt to avoid the melting of the glass phase and reactions between film and substrate material. To produce a single-phase barium hexaferrite film, this would require further optimization. For example, the diffusion of the barium into the substrate needs to be compensated. Since, as discussed above, this substrate material shows lower loss at high frequencies than the DuPont 951, further work on this topic may still be worthwhile to produce low loss components.

BaM film on Heraeus CT 702

SEM images of the film produced on the Heraeus CT 702 tape are shown in Figure 3.31. On the overview in Figure 3.31(a), it can be seen that the film contains pores, but no cracks are observed except for regions close to the substrate edges and corners. The more detailed view of the surface in Figure 3.31(b) shows that the film has a rather low density and a smaller crystal size in comparison to the film produced on the DuPont 951 tape shown in Figure 3.26.

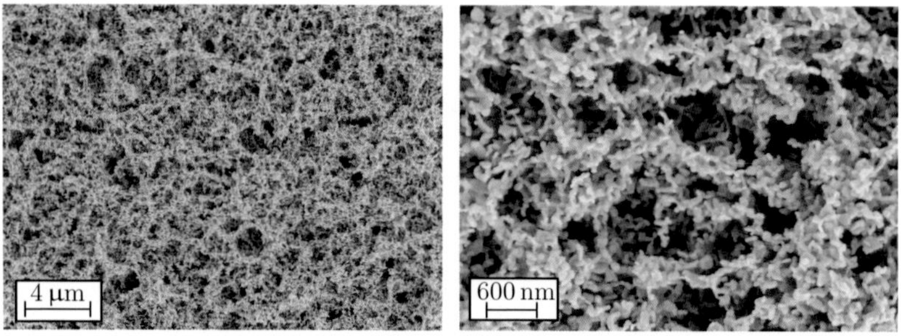

(a) (b)

Figure 3.31: SEM images of a ferrite film produced on Heraeus CT 702 tape.

A longer sintering time could possibly lead to a larger crystal size and increased density, but no further sample was produced due to a lack of additional tape material.

The result of an XRD-analysis performed on the film with the CT 702 as substrate material is depicted in Figure 3.32. Hematite, barium monoferrite, or other secondary phases are not detected in this sample. For nano scale particles, which are smaller than $100-200\,\text{nm}$, it is possible to determine a lower bound for the ordered crystalline domain or grain size. The Scherrer equation [78] can be used to determine the grain size

$$\tau = \frac{K \cdot \lambda}{\beta \cdot cos\Theta}, \tag{3.4}$$

with K the shape factor of 0.9, λ the wavelength of the used Cu K-α radiation, β the line broadening at half maximum, and the Bragg angle Θ. It yields a lower bound for the grain size of $31.2\,\text{nm}$, which is considerably smaller than the grain size observed before, but matches well with what can be determined from the SEM image.

From these results, it is concluded that the production of a barium hexaferrite film with the CT 702 tape as a substrate material is possible. The produced film shows pores and is presumed to have a low density, but no secondary phases are detected. Furthermore, the film is crack-free and has good adhesion to the substrate. In further studies, the effect of a longer sintering time on the produced film could be investigated.

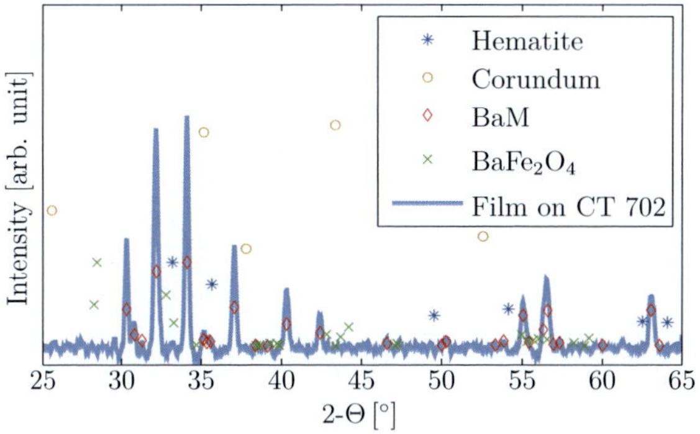

Figure 3.32: XRD-analysis result of a ferrite film prepared on Heraeus CT 702 LTCC tape.

BaM film on Ferro A6M

The surface of a film with the Ferro A6M tape as a substrate is shown in Figure 3.33. In the overview SEM image in 3.33(a), it can be seen that the sample surface again looks fundamentally different from the ones observed before. Furthermore, a rather large vertical crack can be seen in Figure 3.33(a). Cracks and pores of various sizes can indeed be found on the whole sample. Figure 3.33(b) shows a more detailed view of the film surface, on which no BaM crystallites can be identified. The fundamentally different surface properties could again be due to the residual glass phase in the GC tape prior to the post-firing step. The softening point of the borosilicate glass remaining in the substrate after the initial sintering is close to 820 °C [74]. Macroscopically, the film does not look different from films produced on alumina substrate, but it shows a rather bad adhesion. Due to the large cracks, the film can easily be peeled off the substrate.

Figure 3.34 shows the XRD-analysis result for the film on Ferro A6M substrate. Only the largest peaks expected for barium hexaferrite can be identified. The major phase in this sample is hematite which can be identified by two peaks at 33.2 ° and 35.7 °. As discussed before, this GC tape contains no crystalline filler particles. This is in contrast to all the other tested LTCC tapes which contain alumina and possibly a reason for the difficulties in producing a BaM film on this substrate material.

(a) (b)

Figure 3.33: SEM images of a ferrite film produced on Ferro A6M tape.

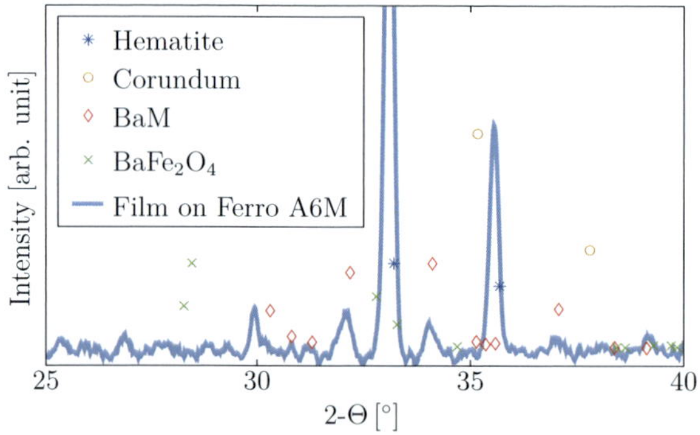

Figure 3.34: XRD-analysis result for a ferrite film produced on Ferro A6M LTCC tape.

In this work, no barium hexaferrite film on the Ferro A6M substrate could be produced. A lower sintering temperature below the melting point of borosilicate glass could possibly reduce the interaction between the substrate and the applied film. Since the sintering temperature for the BaM film needs to be reduced, to accomplish this further time-consuming optimization with a rather uncertain outcome would be required. Again, as for the DuPont 9k7 tape, this further work might be worthwhile. As discussed above, the Ferro A6M tape shows the lowest loss of the tested LTCC substrate materials. It would therefore be a suitable material to produce low-loss components working at high frequencies.

4 Self-Biased Circulator

This chapter is focused on self-biased circulators based on the hard magnetic ferrite films produced by means of the sol-gel method. For this, the basic operation of y-junction circulators is briefly explained and finally the influence of some material and design parameters on the circulator performance are analyzed.

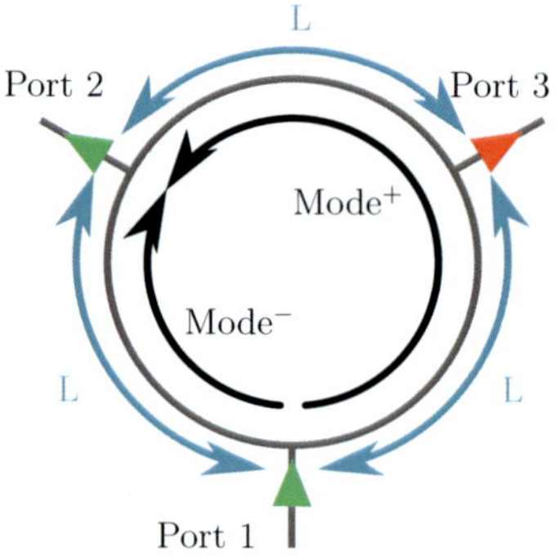

Figure 4.1: Rotating modes in a three port y-junction circulator.

As depicted in Figure 4.1, the basic operation principle of y-junction circulators is based on two modes rotating in opposite directions. In the shown case, transmission from port 1 to port 2 is intended, with port 3 being isolated. For this to occur, the clockwise mode (−) and the counterclockwise mode (+) must have a different phase velocity, yielding a phase difference of $N2\pi$ at port 2. This results in the condition

$$2\beta^+L - \beta^-L = 2N\pi \tag{4.1}$$

with $\beta^{\pm}$ the phase constants of the two modes, $N \geq 1$ an integer, and L the distance between the two ports. To achieve isolation of port 3, the two waves must cancel out giving

$$-\beta^{+}L + 2\beta^{-}L = (2M-1)\pi \tag{4.2}$$

with M being an integer. Solving (4.1) and (4.2) yields

$$\beta^{+}L = \frac{4N + 2M - 1}{3}\pi \tag{4.3}$$

and

$$\beta^{-}L = \frac{2N + 4M - 2}{3}\pi. \tag{4.4}$$

If these two conditions are satisfied simultaneously and losses are omitted, perfect circulation is achieved [2]. The phase constants of the two rotating modes are given by

$$\beta^{-} = \omega\sqrt{\epsilon\frac{\mu^2 - \kappa^2}{\mu}} \tag{4.5}$$

and

$$\beta^{+} = \omega\sqrt{\mu_0\epsilon} \tag{4.6}$$

with μ and κ being the components of the permeability tensor derived in Section 2.2.

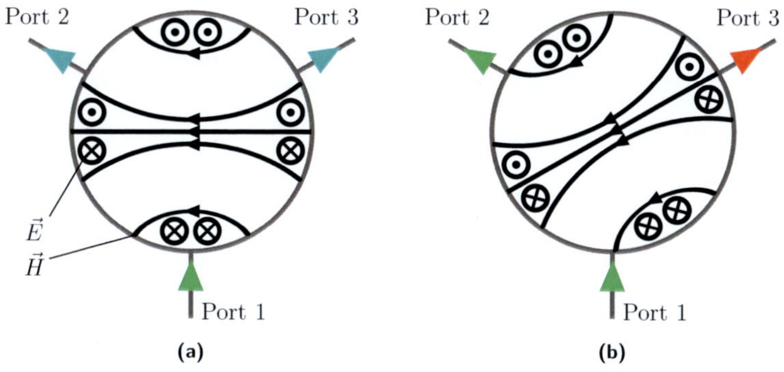

Figure 4.2: Standing-wave pattern in the junction of a circulator (a) without magnetic bias field and (b) with magnetic bias.

Figure 4.2 shows the standing-wave pattern in a circular circulator junction with and without a magnetic bias field in the junction area. Without the magnetic bias field the two counterrotating modes have the same resonant frequency and, as depicted in Figure 4.2(a), the coupling to the two output ports is equal. If a magnetic bias field normal to the plane of the junction is applied, the phase constants and thus the resonant frequencies of the two modes are not the same. Therefore, the standing-wave pattern is rotated, as depicted in Figure 4.2(b), and the coupling to output ports 2 and three is not the same anymore. With $|E| = 0$ at port 3, perfect isolation is achieved. The operating frequency of a circulator is between the resonant frequencies of the counterrotating modes in the circulator junction. However, instead of applying a magnetic bias field to the junction area, a hard magnetic ferrite material is used here. Self-biased circulators based on a variety of hard magnetic materials and fabrication techniques have been presented in other publications. In the following, a brief overview is given.

A self-biased circulator with very good performance is presented in [79]. It uses textured, lanthanum-cobalt-substituted SrM pucks and is built in waveguide technology. At 41.4 GHz it shows an isolation of 28.1 dB and an insertion loss of 1.79 dB, while having a −15 dB bandwidth of 3 GHz. Isolation is defined as the difference between backward and forward transmission ($|S_{12}|/|S_{21}|$ in linear scale). In [80], measurement results of the circulator with a small applied external magnetic biasing field are presented. With the external field, the insertion loss decreases even further to 0.84 dB. An SrM ceramic die with a thickness of 130 μm is used in [81] to build a self-biased microstrip circulator. The die is bonded on a silicon substrate and matching networks are implemented on alumina substrate. The circulator shows an isolation of 37 dB and an insertion loss of 6.9 dB at 28.9 GHz. It has a 1 % bandwidth of 20 dB isolation. In [82], a textured SrM pill is inserted into an alumina substrate with a thickness of 500 μm and a microstrip circulator is built. The device is tuned externally with waveguide tuners to achieve an isolation of more than 50 dB and an insertion loss of 3.5 dB at 23.45 GHz with 5 % bandwidth. In this case, the above resonant operating frequency is chosen based on the available laboratory equipment.

Self-biased circulators using hard magnetic films instead of pills, pucks, or dies show an inferior performance compared to the ones discussed above. A circulator using a sputter-deposited BaM film with a thickness of 10 μm is presented in [19]. The film is single-phase BaM with randomly oriented grains and a grain size of 30 nm. It is deposited on an alumina substrate with the ground plane of the coplanar circulator directly underneath the film on the upside of the substrate. On top of the BaM film, the CPW is patterned in a chromium-gold multilayer which is deposited by evaporation. In measurements, the produced self-biased circulator shows an isolation of 2.6 dB and an insertion loss of 17 dB at 40 GHz.

In [83], another self-biased circulator relying on a sputter-deposited film with preferred c-axis orientation is presented. The BaM film with a thickness of 8 μm is deposited on a sapphire substrate and annealed at 1000 °C. Prior to film deposition, a hole is produced by laser ablation on the backside of the substrate, which is metalized and connected to ground. This increases the ratio of BaM to substrate in the circulator region, which is an important factor in achieving a good circulator performance. A 3 μm thick gold layer is deposited on top of the ferrite film. It is patterned to form the microstrip circulator. In measurements, maximum isolation and minimum insertion loss occur at different frequencies with a maximum isolation of 10 dB at 26.4 GHz and a minimum insertion loss of 27 dB at 25.7 GHz.

A self-biased circulator based on a film of a textured BaM composite magnetic material with a thickness of 100 μm is presented in [28]. The film is produced by spin coating a mixture of photoresist and BaM powder between sputter-deposited copper ground plane and a CPW circulator on an alumina substrate. In the measurements, the self-biased circulator shows an isolation of 1.5 dB and an insertion loss of 6.9 dB at 39.34 GHz.

The mentioned works and two additional ones are compared in Figure 4.3 with regard to the achieved isolation and insertion loss. The circulators in the additional works [84], [85] are based on a SrM disc and substrate, respectively. Only circulators without any externally applied bias field are compared. The result shows that the circulators based on hard magnetic ferrite films have lower isolation and tend to have higher insertion loss in comparison to the ones based on ferrite materials with higher thicknesses. This will be further discussed in Section 4.5. Nevertheless, ferrite films enable simple integration of nonreciprocal components in the future and are therefore of interest.

In this work, a self-biased circulator based on the presented BaM films is fabricated. As already discussed, the films are untextured and have a thickness close to 10 μm. Due to limited laboratory equipment and problems with the deposition of a copper layer on the ferrite film by a chemical route, it was not possible to deposit a patterned metal layer

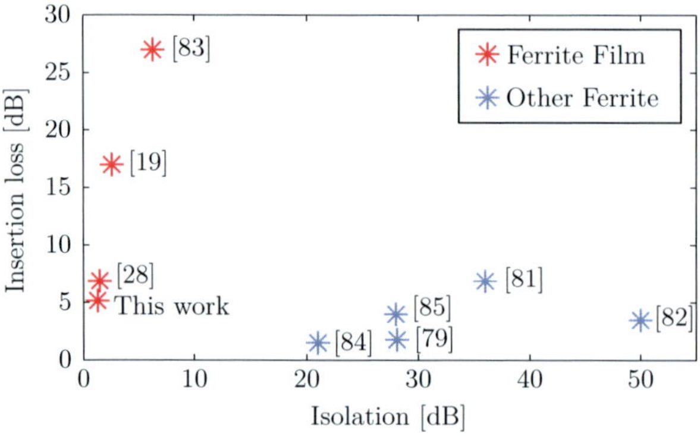

Figure 4.3: State-of-the-art self-biased circulators regarding isolation, insertion loss, and form of the used ferrite material, with "Other Ferrite" relating to pill, puck, die or bulk substrate material.

on the produced ferrite film. A rather thin alumina substrate (127 µm) was used for the circulator design. As shown in Section 3.6, the alumina substrate can be substituted with either DuPont 951 or Heraeus CT702 LTCC tape. With the available equipment, it was also impossible to increase the ratio of ferrite film to substrate thickness in the circulator by thinning the alumina substrate.

4.1 Circulator Concept

In the context of this work, different types of circulators based on the produced BaM films are simulated and some of them are fabricated and measured. The most promising candidates are a CPW circulator and a microstrip y-junction circulator. For the coplanar circulator, an alumina substrate with applied ferrite film is used. Metal layers are needed on both sides, with the coplanar conductors and the circular center conductor of the circulator on the top side and a circular ground plane on the backside. This stack up requires a patterned metalization on the ferrite film. In an attempt to achieve this, the gold on the top side of the alumina substrate was patterned and the ferrite film was deposited on the backside. In a next step, the gold layer was protected by a kapton tape

(a) Backside.

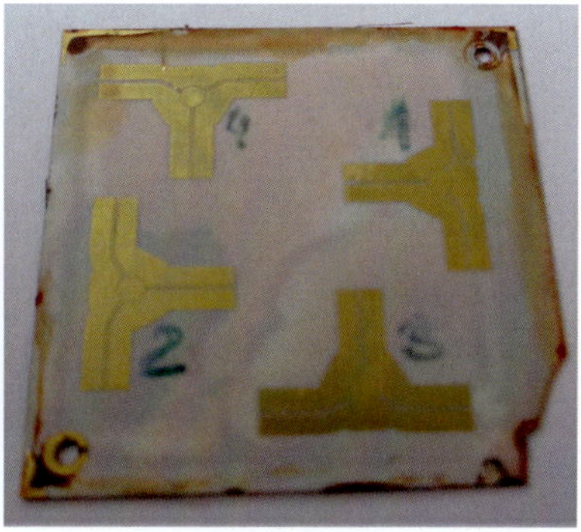

(b) Top side.

Figure 4.4: Photograph of a CPW circulator showing the manufacturing problems.

while copper was deposited chemically on the ferrite film. In a last step, the deposited copper film was patterned.

Figure 4.4 shows a photograph of the top and the backside of such a circulator. This circulator approach produced multiple problems during fabrication, including, as can be seen in Figure 4.4(a), the partial removal of the ferrite during the substrate surface preparation for the copper deposition or during etching of the copper film. Another challenge was to achieve an exact alignment of top and back metallization without the aid of alignment holes which could not reliably be manufactured in alumina. Due to these problems, the coplanar circulator was not further investigated. Instead, a microstrip circulator was designed, built, and measured. Nevertheless, it should be pointed out that the coplanar circulator shows promising behavior in simulation and, for one fortunate production attempt, a higher nonreciprocal effect than the microstrip circulator presented in this thesis. However, due to the various manufacturing problems, this result was not reproducible. Therefore, further work on this circulator, with better laboratory equipment and other metalization technologies, could be promising.

In the following, the design, fabrication, and measurements of a self-biased microstrip y-junction circulator based on the previously discussed BaM films are presented [86]. The main advantage of the microstrip circulator in comparison to the coplanar one is that a

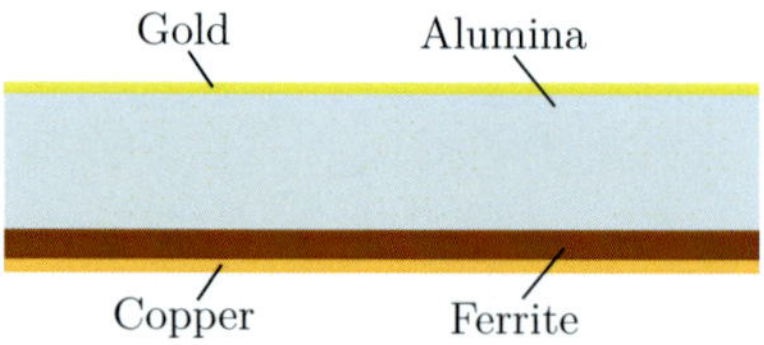

Figure 4.5: Cross-sectional view of the circulator with BaM film and copper ground plane [86].

continuous instead of a patterned metal plane is needed on the ferrite film. Thus, no further etching and alignment steps are necessary.

Figure 4.5 shows a cross-sectional view of the circulator. An alimuna substrate with a thickness of 127 µm is used and the microstrip circulator is patterned on the top side in the 5 µm thick gold layer. The ferrite film with a thickness of 10 µm is deposited as described in Section 2.3.2 on the backside of the substrate and covers the complete substrate surface area. With this design, there is no need to etch the ferrite with hydrochloric acid. This is an advantage since the hydrochloric acid, as described in Section 2.3.2, lowers the adhesion of the gold conductors to the substrate. The copper metalization on the ferrite layer is achieved with an adhesive copper foil covering the backside of the produced circulator and providing a ground plane for the microstrip conductors.

For the design of the microstrip circulator, it is reasonable to first look at the well-known stripline circulator description and design by Bosma [87]. The calculated dimensions can later be used as a starting point in simulation to design a microstrip circulator using a ferrite film. The radius R of the junction area and the required magnet size of a stripline circulator can be calculated as [87]

$$R = \frac{1.84\lambda}{2\pi\sqrt{\mu_{\text{eff}}\epsilon}}, \tag{4.7}$$

with

$$\mu_{\text{eff}} = \frac{\mu^2 - \kappa^2}{\mu}, \tag{4.8}$$

where μ and κ are given in (2.21) and (2.22) and $\omega\sqrt{\mu_0\epsilon_0} = \frac{2\pi}{\lambda}$. In a microstrip circulator, there is no clearly defined edge of the junction area. In [2], a reduction of R is suggested,

Table 4.1: Adjusted material parameters used for the BaM film in the Ansys EM simulation.

Barium hexaferrite film	
Relative dielectric constant	14.2
Dielectric loss tangent	0.01
Saturation magnetization	240 mT
Internal field	970.8 $\frac{\mathrm{kA}}{\mathrm{m}}$
Damping factor	0.02

yielding

$$R = \frac{1.65\lambda}{2\pi\sqrt{\mu_{\text{eff}}\epsilon}}. \tag{4.9}$$

The radius of the center conductor disc of the microstrip circulator is smaller than the junction area and magnet size. As suggested in [2], it is assessed to be

$$r = 0.8R\,. \tag{4.10}$$

4.2 Circulator Design

The circulator is analyzed and optimized in the 3D-full wave simulation software Ansys EM 2015.2 by Ansys. The material parameters used for the BaM film in the simulation are given in Table 4.1. These parameters are different from those found in literature [13]. In an attempt to account for the remanent magnetization state of the ferrite, the material properties are adjusted. The Polder model [14] used by Ansys EM is valid only for ferrite materials magnetized to saturation [88]. With a modification of the static parameters, the Polder model can yield approximate values for the dynamic permeability [79]. Since the circulator is self-biased and the ferrite material is therefore not magnetized to saturation, the modified material parameters are used in the simulation. With these parameters, (4.10) yields a center disc radius of 697.9 µm, which is used as a starting point for the optimization.

The top view of the circulator is depicted in Figure 4.6. Optimization goals are a high NR effect and a low input reflection. The dimensions of the final circulator are given in Table 4.2 and the center disc radius after optimization is smaller than suggested by (4.10). To

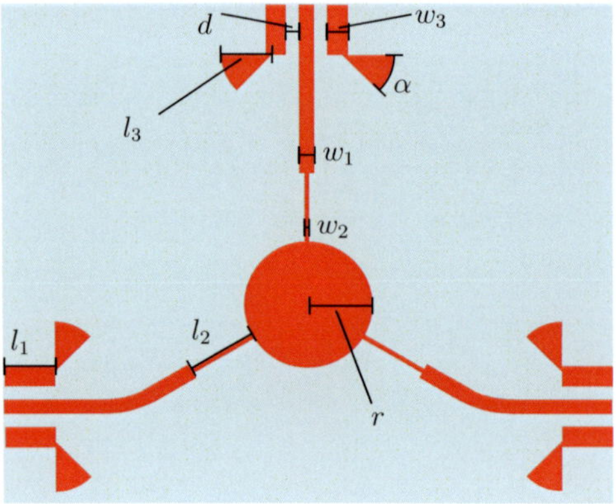

Figure 4.6: Top view of the circulator with two 30° bends and transitions to CPW for measurement probes [86].

Dimensions

r	655 µm
d	130 µm
l_1	500 µm
l_2	686 µm
l_3	500 µm
w_1	143 µm
w_2	30 µm
w_3	200 µm
α	45°

Table 4.2: Final dimensions of the circulator optimized in Ansys EM [86].

simplify the measurement with the available test equipment, 30° bends are attached to two of the three ports, resulting in a 90° angle between the ports.

A transition from CPW to microstrip line is added to each port to enable measurement with the ground-signal-ground probes. For ease of fabrication, a transition needing no via holes [89] is designed and simulated. The dimensions of the transition are also given in Table 4.2. The simulation results for a single transition are depicted in Figure 4.7. From the simulation result it can be seen that the transition has an insertion loss of less than 0.15 dB and an input reflection below −35 dB in the relevant frequency range from 40 GHz to 50 GHz.

The center disc with a radius of 655 µm is connected to the microstrip line via lines with a width of 30 µm. These thin lines increase the isolation as well as the insertion loss of the circulator. A line width of 30 µm is close to the minimum for reliable fabrication. For a first component, the higher loss is accepted in exchange for the benefit of increased isolation. The length of these lines is adjusted to minimize the input reflections.

Figure 4.8 shows the simulation results for the circulator. The input reflection is less than −10 dB from 41.8 GHz to 43.3 GHz. The minimum insertion loss in the preferred

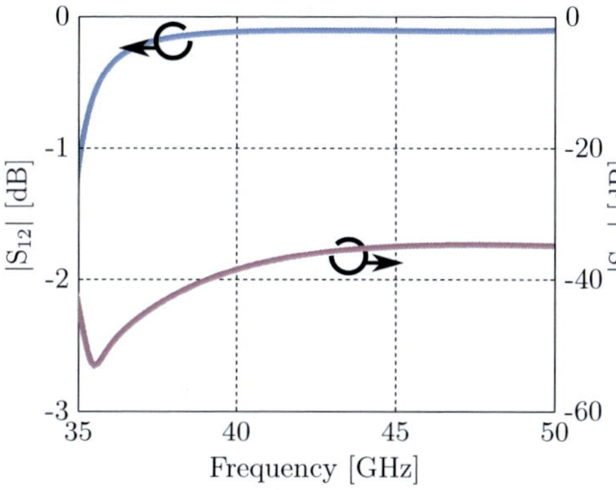

Figure 4.7: S-parameter simulation results of the transition from CPW to microstrip line.

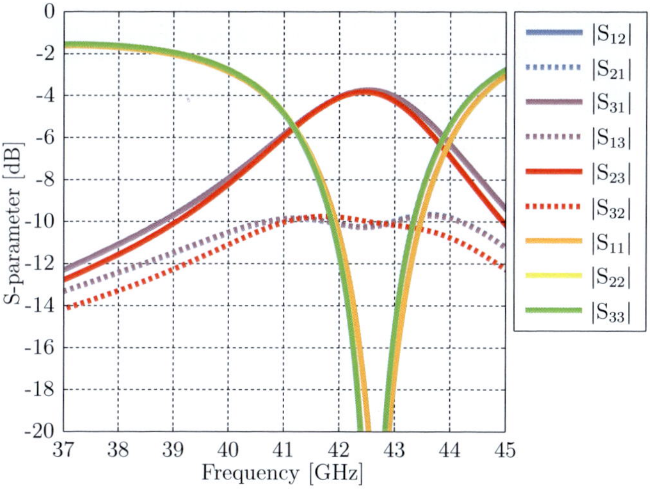

Figure 4.8: Simulated S-parameters of the self-biased circulator [86].

transmission direction of the circulator is 3.66 dB and thus quite high. This is partly caused by the very thin lines connected to the center disc. The transmission in reverse direction is −9.8 dB at 42.5 GHz and therefore the isolation (NR) at this frequency is 6.14 dB. The simulation results are not completely symmetric for the three ports due to the 30 ° bends on two of the ports.

4.3 Fabrication and Measurements

As the first step in fabrication, the designed circulator is patterned on the top side of a gold-coated alumina substrate. Afterwards, the ferrite film is produced as described in Section 2.3.2. Again, 20 layers are deposited in order to achieve a thickness close to 10 μm. Each individual layer is dried at 550 °C and the film is subsequently sintered at 850 °C for 3 h. In the last step, the self-adhesive copper foil is attached to the ferrite film.

(a) Ferrite film deposited on the backside of the alumina substrate.

(b) Image of the backside with the copper foil on top of the ferrite film.

Figure 4.9: Photographs of the backside of the circulator.

Figure 4.9 shows images of the backside of the circulator. Figure 4.9(a) shows the deposited BaM film after sintering. It can be seen that the film surface is quite homogeneous except for the substrate edges and corners. Since the circulator is placed closer to the center, these regions are not of importance. Figure 4.9(b) shows the backside of the circulator with copper foil, which functions as a ground plane for the microstrip line. Prior to the measurements, the film is magnetized to saturation using an electromagnet. The circulator during measurements is shown in Figure 4.10, whereas Figure 4.11 shows the measured S-parameters of the realized circulator. Three-port measurements are performed using a

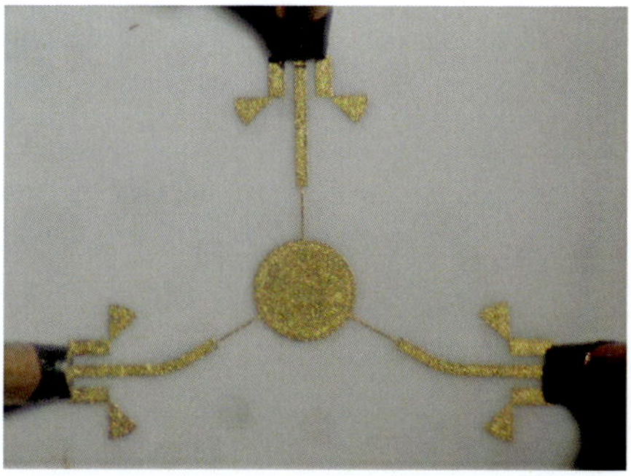

Figure 4.10: Image of the top side of the circulator during measurement [86].

four-port vector network analyzer. During the measurements, the ferrite is in the remanent magnetization state, which means that no external magnetic bias field is applied.

Figure 4.11 depicts the measurement results for the self-biased circulator. The operating frequency at which the input reflection is minimum differs from the simulated one. In the measurements it is close to 47 GHz. The 10 dB bandwidth is 1.86 GHz and the minimum insertion loss of 5.17 dB is high in comparison to the simulation. The minimum input reflection is highest at port 3 (-14.8 dB). A possible reason for the frequency shift is the shrinkage of the gold conductors, which occurred during the sintering of the BaM film at 850 °C. Possible reasons for the asymmetry of the results are the bends at two of the ports and tolerances in the fabrication process.

The NR effect is difficult to assess from Figure 4.11 and is therefore plotted separately in Figure 4.12. The measured NR is close to 1.3 dB at 47 GHz. This effect is smaller than predicted by simulation and far from the performance of commercially available, conventionally fabricated circulators. However, the preferred transmission direction can be seen clearly for all three ports and this measurement result is, as can be seen in Figure 4.3, comparable to results obtained by others with BaM films for self-biased circulators.

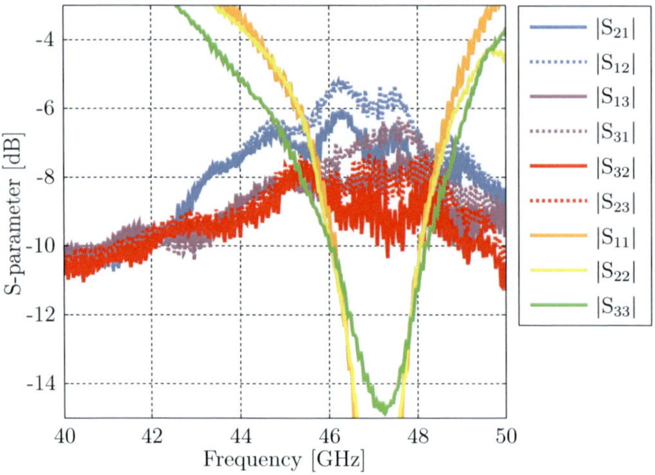

Figure 4.11: Measured S-parameters of the self-biased circulator [86].

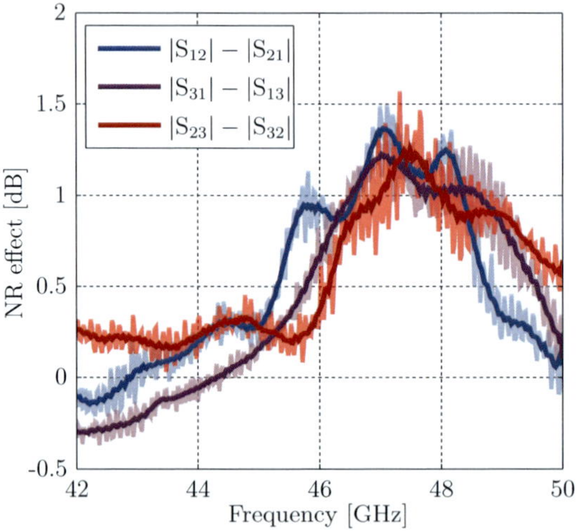

Figure 4.12: Measured NR effect of the self-biased circulator [86].

4.4 Circulator with Aluminum-Doped Ferrite

A self-biased circulator based on a doped barium hexaferrite film is designed, fabricated, and measured. In comparison with the scandium-doped films, the aluminum-doped ones show a higher remanent magnetization. Therefore, an aluminum-doped film with a doping factor of $y = 0.8$ is chosen. According to the previous experiments, this film exhibits its FMR at 61.7 GHz.

The circulator concept and design is similar to the one used in Section 4.2, but is optimized for operation at this higher frequency. For the measurements in this higher frequency range, a two-port network analyzer is used. Since only two ports are available, the third port is matched and the behavior of an isolator is expected. Reflections at the matched port lower the isolation and therefore need to be avoided. To achieve a low reflection, the resistive paste SD2843HAL supplied by Peters Lackwerke is used for the fabrication of a load. It is coated on straight microstrip lines of 5 mm and 10 mm in length using a doctor knife. To control the coated area and thickness, a paper with cutouts and a thickness of 37 µm is used. The produced film is dried at 150 °C for 30 min. and the thickness of the final film is measured to be between 7 µm and 17 µm.

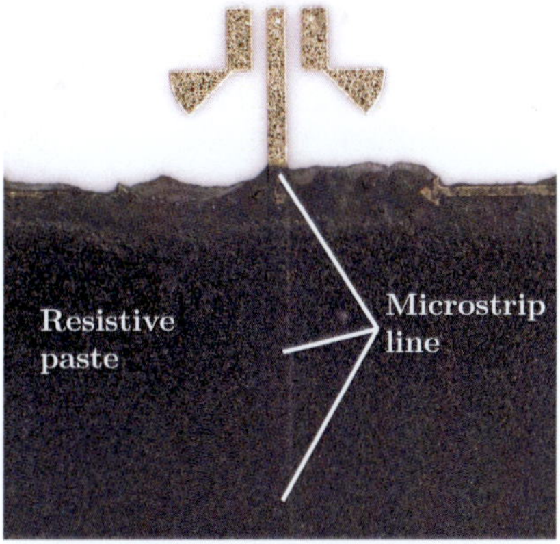

(a) Picture of the load.

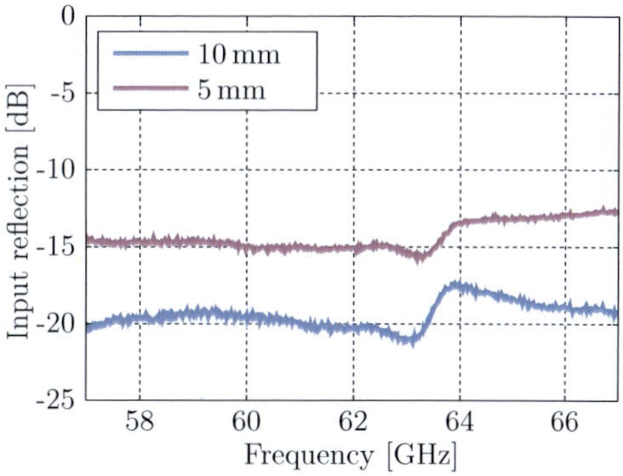

(b) Measured input reflection of loads with lengths of 10 mm and 5 mm.

Figure 4.13: Top view of the load and measurement results for loads with different lengths.

Parameters	
Internal field	$1372\,\frac{\text{kA}}{\text{m}}$
r	487 µm
l_2	448 µm
w_2	47 µm

Table 4.3: Changed parameters compared to circulator with the undoped ferrite film.

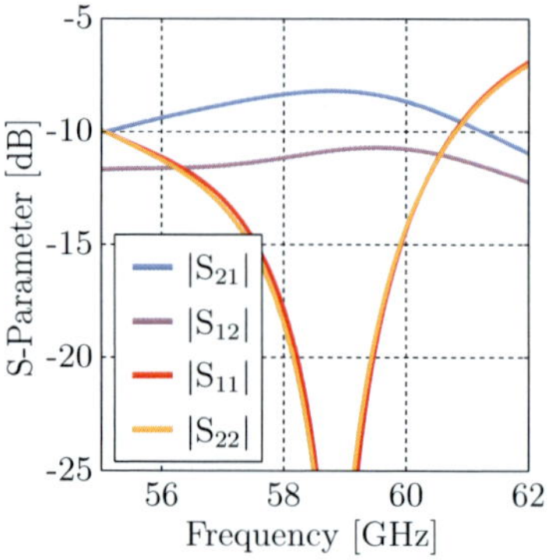

Figure 4.14: Simulated S-parameters of the circulator with aluminum-doped ferrite film.

Figure 4.13(a) shows a photograph of a load that is fabricated separately for testing. A transition to coplanar waveguide is used to enable measurements with coplanar probes. A calibration with the reference plane behind the transition is performed prior to the measurement. The black resistive paste covers the straight, open-ended microstrip line. Two lengths, namely 5 mm and 10 mm, for the covered part of the microstrip line are tested. Figure 4.13(b) shows the measurement results for the input reflection of these loads. Both have an input reflection of less than −10 dB over a broad bandwidth which justifies this simple approach. The input reflection of the 10 mm-long microstrip line is lower than −17 dB in the frequency range of interest and is thus chosen for the termination of the third circulator port.

The design of the circulator based on an aluminum-doped ferrite film is similar to the one with the undoped film shown in Figure 4.6. The parameters that are changed are summarized in Table 4.3. Figure 4.14 shows the simulation results for the circulator using the aluminum-doped ferrite film. The input reflection is below −10 dB from 55.7 GHz to 60.9 GHz. The minimum insertion loss is 8.18 dB at 58.5 GHz and the isolation (NR) at the same frequency is 2.74 dB.

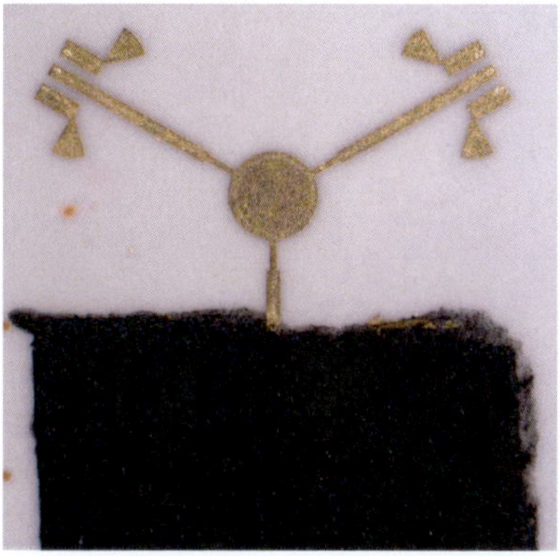

Figure 4.15: Picture of the measured circulator based on the doped BaM film with one port terminated by a load.

Figure 4.15 shows the fabricated circulator. Since no 30 ° bends are employed, the two ports used for the measurement form an angle of 120 °. Therefore, a fixture is fabricated that enables the measurement to be performed with probes tilted by an angle of 30 °. The third port is terminated as described above with a load of 10 mm length. The resistive paste is applied as a last step after sintering the ferrite film and attaching the copper ground foil. The paste is therefore not exposed to temperatures higher than 150 °C. The circulator performance is optimized with regard to high isolation using simulation. As in the design of the circulator with the undoped film described in Section 4.2, thin microstrip lines are used at the center disc to achieve a higher isolation.

The results of the two-port measurements are displayed in Figure 4.16. The input reflection $|S_{11}|$ shown in Figure 4.16(a) is lower than −10 dB between 61 GHz and 63.3 GHz. At 63 GHz, the frequency with the maximum NR effect, the insertion loss is 7.7 dB. The former is depicted separately in Figure 4.16(b). It is rather low with a maximum of 0.31 dB at 63 GHz. As for the circulator based on the undoped film, this value is lower than expected from simulation.

The operating frequency of the circulator is close to the FMRF of the chosen aluminum-doped BaM ferrite at 61.7 GHz. This is a possible reason for the high insertion loss, but the NR effect is nevertheless rather small. The small NR could be caused by a reduction

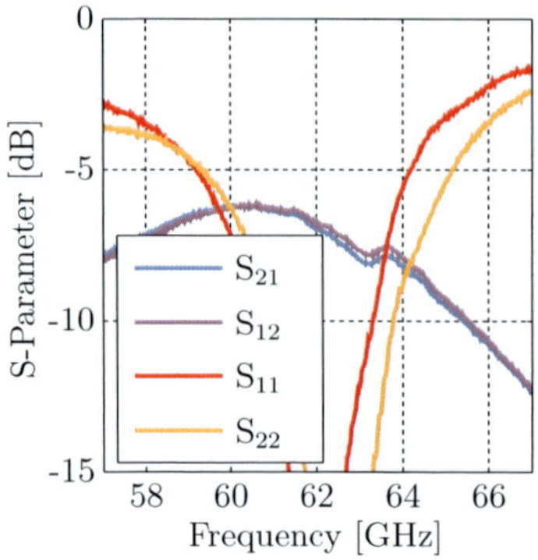

(a) S-parameter measurement result.

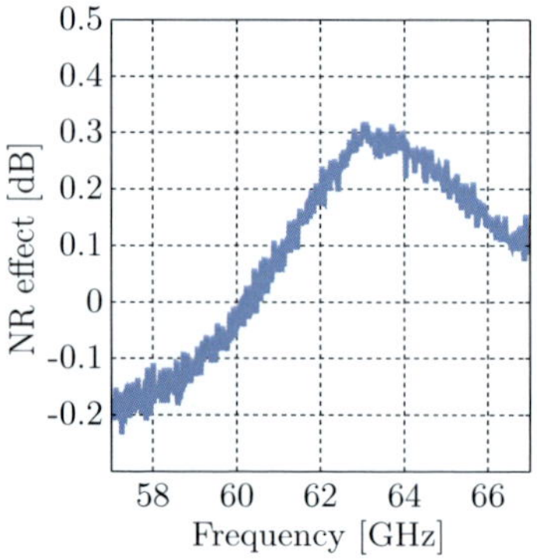

(b) Measured NR effect.

Figure 4.16: Measurement results of the circulator based on an aluminum-doped BaM ferrite film.

of the remanent magnetization of the material due to the aluminum doping. The VSM-measurement results presented in Figure 3.23 (Section 3.4.2), show that the reduction of the magnetization due to the doping is less for the aluminum-doped films compared to the scandium-doped ones. The material with $y = 0.5$ exhibited 77 % of the magnetization of the undoped film. An even further reduction due to the higher doping factor of $y = 0.8$ could be responsible for the rather small NR effect of 0.31 dB.

These first measurement results show that it is in principle possible to build a self-biased nonreciprocal component using aluminum-doped BaM films to enable higher operating frequencies. Improvements of the ferrite material produced in this work, which are necessary to enhance the performance of the circulators based on this material, are discussed in Section 4.5.

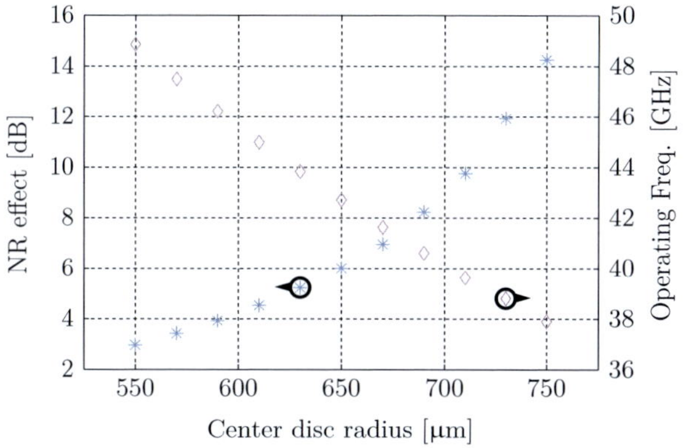

Figure 4.17: Simulated NR effect and operating frequencies for circulators with different center disc radii [86].

4.5 Discussion of Results

In this discussion, the circulator depending on the undoped BaM film is used, but the challenges are similar for the circulator based on the doped material. The circulators described at the beginning of Chapter 4 are based on BaM thin films produced by different technologies [19], [28], [83]. Their performance is comparable to the measurement result in this work. The achieved NR effect ranges from 1.5 dB to 6.3 dB and the insertion loss in these works is between 6.9 dB and 27 dB. To be comparable with commercially available waveguide circulators in the same frequency band, an improved performance of the components using hard magnetic films is necessary. There are multiple challenges in the design of self-biased circulators based on a hard magnetic films.

For conventional nonreciprocal components, the external bias field can be tuned to adjust the FMRF. For self-biased components, the anisotropy field is determined by the material and cannot be adjusted and tuned during measurements. The operating frequency has to be chosen carefully, since losses in the ferrite material are high close to the FMRF. Therefore, a circulator with an operating frequency too close to the FMRF would exhibit a high insertion loss.

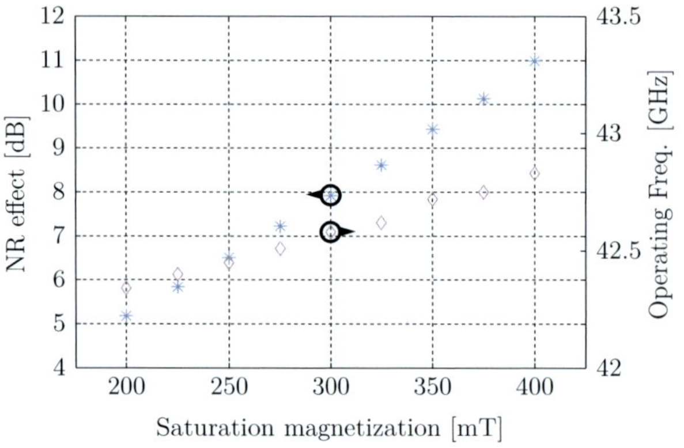

Figure 4.18: Simulated influence of the ferrite magnetization (saturation magnetization in the Polder model) on the NR effect [86].

To investigate the influence of the operating frequency on the nonreciprocal effect, the center disc radius of the circulator (r in Figure 4.6) is varied in simulation, whereas all other dimensions and material parameters are kept constant. Figure 4.17 shows the maximum NR effect and the corresponding operating frequency for center disc radii from 550 μm to 750 μm. The NR effect increases with the radius of the center disc and, therefore, with lower operating frequencies. For the larger radii the operating frequency approaches the FMRF and the NR effect and insertion loss increase. The intended and simulated operating frequency of the circulator is 42.5 GHz. From the simulation results in Figure 4.17 it can be seen that the shift of operating frequency to 47 GHz observed in the measurement, is possibly one of the reasons for the low NR effect of the fabricated circulator.

The static magnetic properties of the ferrite material also offer opportunities for improvement. The remanent magnetization of the material is important for the performance of a self-biased circulator. As discussed in Section 3.5, a textured material offers a higher remanent magnetization and is thus used for the circulators based on BaM films in [28] and [83]. Since Ansys EM uses the Polder model, it is not possible to simulate ferrites in the remanent magnetization state and it is reasonable to enter an adjusted saturation magnetization. This value is 240 mT for the BaM films produced in this work.

To analyze the influence of this adjusted saturation magnetization on the NR effect of the circulator, simulation runs are performed in which this value is varied, whereas all other material parameters and dimensions are kept constant. The results are depicted in Figure 4.18. It can be seen that the NR effect increases with increasing saturation magnetization of the ferrite film. Also, the operating frequency is influenced. However, this frequency shift is not the cause of the improved NR effect, since the operating frequency increases with higher magnetization. According to the results in Figure 4.17, the higher operating frequency actually decreases the NR effect.

The simulated results presented in Figure 4.18 show the importance of producing a textured ferrite film. In [13], a value of 480 mT is given for the saturation magnetization of BaM bulk material. Textured films with high squarenesses of up to 0.95 have been demonstrated in the past [64]. The saturation magnetization reported for BaM films is consistently below the value for bulk material. However, the film in [64] with $M_S = 400\,\mathrm{mT}$ and a squareness of 0.95, for example, offers a remanent magnetization of 380 mT. In comparison, the films in this work are simulated with a magnetization of 240 mT. If the challenges discussed in Section 3.5 can be overcome and a BaM film with high squareness and 10 µm thickness can be produced, the performance of self-biased components could be significantly improved.

The saturation and remanent magnetization of the produced film are also lowered by nonmagnetic inclusions in the material. The density of the films produced in this work was not determined, but the SEM images suggest a rather low density. Air inclusions in the film can therefore be another reason for the realized low NR effect. In this work, the formation of large pores was avoided by adjusting the drying temperature of the film. The density of the produced ferrite film was not further analyzed or optimized. Therefore, it is quite possible that a higher saturation magnetization can be reached by further optimizing the process parameters to achieve a higher density. Also, the formation of nonmagnetic secondary phases needs to be avoided to achieve a high saturation magnetization of the produced ferrite film. In this work, this is achieved by avoiding the formation of, for example, hematite and barium monoferrite as discussed in Section 3.2.

The ratio of the ferrite material thickness to the overall height of the circulator junction is called filling factor in the following and has a significant influence on the circulator performance. For conventional components based on ferrite pills or substrates, the filling factor is usually 1. For the circulator in this work, it is close to 0.073. The filling factor could be

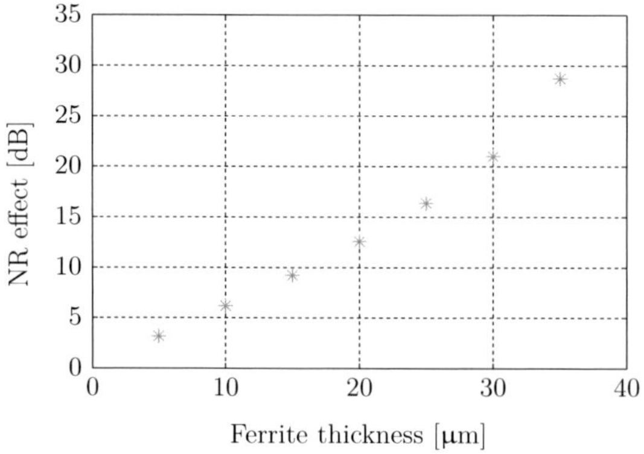

Figure 4.19: The influence of the ferrite film thickness on the NR effect in simulation [86].

improved by using a thicker ferrite film. Simulation results for different film thicknesses are depicted in Figure 4.19. Again, all other material parameters and dimensions are kept constant for this simulation.

The simulated results show that the NR effect can be increased significantly with a larger film thickness. An NR effect of 28.75 dB is achieved with a film thickness of 35 µm, corresponding to a filling factor of 0.21. As discussed in Section 3.1, producing films with a thickness larger than 10 µm proved to be difficult and led to cracking and bad adhesion of the produced film to the substrate. Another possible way to increase the filling factor is to reduce the substrate height. The alumina substrate used for the circulator in this work has a thickness of 127 µm and is already quite fragile. Thin substrates tend to crack on the vacuum chuck of the spin coater. Therefore, using even thinner alumina substrates is not an option. Because it is available with a minimum tape thickness of 40 µm after sintering, the DuPont 951 LTCC tape offers a possible solution to this problem. The combination of this tape with a 10 µm film would result in a filling factor of 0.2. In LTCC, cracking of the substrate could be avoided by using extra tape layers underneath the ground plane to improve the stability. The problem would also be solved by having no substrate in the junction but only the ferrite film, yielding a distance of 10 µm from microstrip line to ground plane. A similar approach is used in [19] with a CPW circulator. However, a relatively high input reflection proved to be a problem.

The simple solution to this problem, a ferrite film with a larger thickness, could possibly be produced by means of the sol-gel method in a clean-room environment. Due to limited equipment, this could not be tested in this work. Since each layer has to be dried for five minutes, producing a thick film is time-consuming if the process is not automated.

5 Conclusion

In this work, barium hexaferrite films for application in small, self-biased nonreciprocal components are fabricated by means of the sol-gel method. As a first step, the fabrication process of the films is analyzed and optimized. This results in a process by which single-phase and crack-free barium hexaferrite films can be produced on ceramic substrates at comparably low temperatures. Measurements on a coplanar waveguide show the ferromagnetic resonance of the material at 43.2 GHz without externally applied bias field. To influence the crystalline anisotropy of the material and therefore its ferromagnetic resonant frequency, the ferrite is doped with scandium or aluminum. By means of doping, the resonant frequency can be shifted over a broad range of frequencies from 27.1 GHz to 65.1 GHz. The scandium-doped films show no secondary phases, but the saturation magnetization of the material is lowered due to the doping. For the aluminum-doped films, the reduction in saturation magnetization is smaller even though the formation of barium monoferrite in these films cannot be prevented. The fabrication of the barium hexaferrite films is transferred from alumina substrates to four different kinds of low-temperature co-fired ceramic (LTCC) tapes. On Dupont 951 and Heraeus CT 702 tape single-phase barium hexaferrite films can be produced. However, the films fabricated on Dupont 9k7 and Ferro A6M exhibit secondary phases which presumably form due to reactions between the film and the substrate during sintering.

Finally, the fabricated material is used to implement a circulator based on these ferrite films. A microstrip y-junction circulator is designed, fabricated, and measured. During the measurement, no external magnetic field is applied. The first results show a promising nonreciprocal effect of 1.3 dB close to 47 GHz with a minimum insertion loss of 5.17 dB. A second circulator using an aluminum-doped film is implemented and exhibits a nonreciprocal effect of 0.31 dB at 63 GHz. The influence of some material and design parameters are analyzed and it is discussed how to improve the performance of circulators based on these hard magnetic ferrite films.

The fabricated barium hexaferrite films could offer simple and cost-effective means of fabricating small and lightweight nonreciprocal components capable of integration into low-temperature co-fired ceramics. The ferromagnetic resonant frequency can be adjusted in a broad frequency range by doping, making the films suitable for various applications. Especially for future satellite communications in the Q-/V-band, small integrated components with low weight are of interest.

To enable self-biased nonreciprocal components based on these films that are able to compete with commercially available components based on soft magnetic ferrites, several material parameters should be improved. It would be most desirable to produce films with a larger thickness, a preferred *c*-axis orientation of the crystallites, and a high density. Furthermore, the component design should be revised to find a favorable circulator concept for the fabricated films. For a new design, the possibility to deposit a conductive layer on the ferrite film could be advantageous. If these challenges can be overcome in the future, these films could enable a new kind of small and lightweight nonreciprocal component.

List of Elements

Ag .... silver

Al .... aluminum

As .... arsenic

Au .... gold

Ba .... barium

C .... carbon

Ca .... calcium

Cl .... chlorine

Co .... cobalt

Cr .... chromium

Cu .... copper

Ga .... gallium

H .... hydrogen

In .... indium

K .... potassium

Mg .... magnesium

Mn .... manganese

N .... nitrogen

Ni .... nickel

Pb .... lead

Pt .... platinum

Ru .... ruthenium

Sc .... scandium

Si .... silicon

Sn .... tin

Sr .... strontium

Ti .... titanium

Zn .... zinc

Zr .... zirconium

List of Abbreviations

BaM .... barium hexaferrite M-type ($BaFe_{12}O_{19}$)

CE .... "glass-free" ceramic

CGS .... *Centimetre Gram Second*

CPW ...... coplanar waveguide

CVD ...... chemical vapour deposition

DSC .... differential scanning calorimetry

EDXS ..... energy-dispersive X-ray spectroscopy

FMR ...... ferromagnetic resonance

FMRF .... ferromagnetic resonant frequency

GBC ...... glass bonded ceramic

GC .... glass ceramics

GCC ...... glass ceramic composite

GID .... grazing incidence diffraction

IP .... in-plane

LPE .... liquid-phase epitaxy

LTCC ..... low temperature co-fired ceramics

MOD ...... metallo-organic decomposition

NR .... nonreciprocal

OP .... out-of-plane

PLD .... pulsed laser deposition

PVP .... polyvinylpyrrolidone

SEM .... scanning electron microscope

SI .... *Système international d'unités (International System of Units)*

SrM .... strontium hexaferrite M-type ($SrFe_{12}O_{19}$)

TGA ...... thermogravimetric analysis

TUHH .... Hamburg University of Technology

UHH ...... University of Hamburg

VSM ...... vibrating sample magnetometer

XRD ...... X-ray diffraction

List of Figures

List of Tables

Bibliography

[1] A. J. Baden Fuller, *Ferrites at Microwave Frequencies.* Peregrinus, London, United Kingdom, 1987.

[2] D. K. Linkhart, *Microwave Circulator Design*, 2nd E. Artech House, London, United Kingdom, 2014.

[3] Y. Imanaka, *Multilayered Low Temperature Cofired Ceramics (LTCC) Technology.* Springer Science & Business Media, Berlin, Germany, Jan. 2005.

[4] C. J. Brinker and G. W. Scherer, *Sol-Gel Science The Physics and Chemistry of Sol-Gel Processing*, G. W. Scherer, Ed. Boston Academic Press, Boston, MA, USA, 1990.

[5] V. G. Harris, "Modern Microwave Ferrites", *IEEE Transactions on Magnetics*, vol. 48, no. 3, pp. 1075–1104, Mar. 2012.

[6] J. Smit and H. Wijn, *Ferrites, Physical Properties of Ferrimagnetic Oxides in Relation to their Technical Applications.* Philips Research Laboratories, Eindhoven, Netherlands, 1959.

[7] D. M. Pozar, *Microwave Engineering*, 4. ed. John Wiley & Sons, Hoboken, NJ, USA, 2012.

[8] Q. Li, X. Guo, H. Song, K. K. Li, Y. K. Zou, H. Jiang, and A. Gopinath, "60GHz Substrate Integrated Waveguide Resonance Isolator", in *2016 IEEE MTT-S International Microwave Symposium (IMS)*, San Francisco, USA, May 2016.

[9] P. B. Braun, *The Crystal Structures of a New Group of Ferromagnetic Compounds*, Philips Res. Rep. 12. Eindhoven, Netherlands, Dec. 1956.

[10] Ü. Özgür, Y. Alivov, and H. Morkoç, "Microwave Ferrites, Part 1: Fundamental Properties", *Journal of Materials Science: Materials in Electronics*, vol. 20, no. 9, pp. 789–834, 2009.

[11] V. Adelsköld, "X-Ray Studies on Magneto-Plumbite, PbO. 6Fe2O3, and Other Substances Resembling "Beta-Alumina", Na2O. 11Al2O3", *Arkiv för Kemi Mineralogioch Geologi, Series A-12*, vol. 29, pp. 1–9, 1938.

[12] M. Wu, *M-Type Barium Hexagonal Ferrite Films*. INTECH Open Access Publisher, Department of Physics, Colorado State University, Fort Collins, USA, 2012, 29 pp.

[13] B. Lax and K. J. Button, *Microwave Ferrites and Ferrimagnetics*, ser. Lincoln Laboratory publication. New York, NY, USA, McGraw-Hill, 1962.

[14] D. Polder, "On the Theory of Ferromagnetic Resonance", *The London, Edinburgh, and Dublin Philosophical Magazine and Journal of Science*, vol. 40, no. 300, pp. 99–115, 1949.

[15] C. Kittel, "On the Theory of Ferromagnetic Resonance Absorption", *Physical Review*, vol. 73, no. 2, p. 155, Jan. 1948.

[16] R. C. Pullar, "Hexagonal Ferrites: A Review of the Synthesis, Properties and Applications of Hexaferrite Ceramics", *Progress in Materials Science*, vol. 57, no. 7, pp. 1191–1334, Sep. 2012.

[17] S. Sakka, *Handbook of Sol-Gel Science and Technology. 1. Sol-Gel Processing*, S. Sakka, Ed. Springer Science & Business Media, 2005, vol. 1.

[18] A.-S. Dehlinger, M. Le Berre, B. Canut, J. Chatelon, D. Albertini, S. Perrot, D. Givord, and J. Rousseau, "Microstructural and Magnetic Properties of Thick ($\geq$ 10µm) Magnetron Sputtered Barium Ferrite Films", *Journal of Magnetism and Magnetic Materials*, vol. 322, no. 21, pp. 3293–3297, Nov. 2010.

[19] A.-S. Dehlinger, M. Le Berre, E. Bénevent, H. Hassane, D. Givord, V. Larrey, and D. Vincent, "Development of Millimeter Wave Integrated Circulator Based on Barium Ferrite", *Materials Science and Engineering: C*, vol. 28, no. 5-6, pp. 755–758, Jul. 2008.

[20] E. J. Donahue and D. M. Schleich, "The Deposition of $BaFe_{12}O_{19}$ by Metalorganic Chemical Vapor Deposition", *Journal of Applied Physics*, vol. 71, no. 12, pp. 6013–6017, Mar. 1992.

[21] S. Pignard, H. Vincent, and J. Senateur, "Epitaxial and Polycrystalline $BaFe_{12}O_{19}$ Thin Films Grown by Chemical Vapour Deposition", *Thin Solid Films*, vol. 350, no. 1, pp. 119–123, Aug. 1999.

[22] C. Carosella, D. Chrisey, P. Lubitz, J. Horwitz, P. Dorsey, R. Seed, and C. Vittoria, "Pulsed Laser Deposition of Epitaxial $BaFe_{12}O_{19}$ Thin Films", *Journal of Applied Physics*, vol. 71, no. 10, pp. 5107–5110, Feb. 1992.

[23] S. Yoon and C. Vittoria, "Microwave and Magnetic Properties of Barium Hexaferrite Films Having the *c*-Axis in the Film Plane by Liquid Phase Epitaxy Technique", *Journal of Applied Physics*, vol. 93, no. 10, pp. 8597–8599, May 2003.

[24] Y.-Y. Song, J. Das, Z. Wang, W. Tong, and C. E. Patton, "In-Plane *c*-Axis Oriented Barium Ferrite Films with Self-Bias and Low Microwave Loss", *Applied Physics Letters*, vol. 93, no. 17, p. 172 503, Oct. 2008.

[25] Y. Nie, I. Harward, K. Balin, A. Beaubien, and Z. Celinski, "Preparation and Characterization of Barium Hexagonal Ferrite Thin Films on a Pt Template", *Journal of Applied Physics*, vol. 107, no. 7, p. 073 903, Apr. 2010.

[26] I. Harward, Y. Nie, A. Gardner, L. Reisman, and Z. Celinski, "Al doped Ba hexaferrite ($BaAl_xFe_{12-x}O_{19}$) thin films on Pt using metallo-organic decomposition", *Journal of Applied Physics*, vol. 111, no. 7, 07A514, Feb. 2012.

[27] Y. Chen, T. Sakai, T. Chen, S. D. Yoon, A. L. Geiler, C. Vittoria, and V. G. Harris, "Oriented Barium Hexaferrite Thick Films with Narrow Ferromagnetic Resonance Linewidth", *Applied Physics Letters*, vol. 88, no. 6, p. 062 516, Feb. 2006.

[28] T. Boyajian, D. Vincent, S. Neveu, M. LeBerre, and J. Rousseau, "Coplanar Circulator Made from Composite Magnetic Material", in *IEEE MTT-S International Microwave Symposium Digest (MTT)*, Baltimore, MD, USA, Jun. 2011.

[29] S. Brosius and A. F. Jacob, "LTCC Material Properties after High Temperature Treatment", in *European Microwave Conference (EuMC)*, Nuremberg, Germany, Oct. 2013, pp. 436–439.

[30] M. Matsumoto, A. Morisako, T. Haeiwa, K. Naruse, and T. Karasawa, "Ba-Ferrite Films Prepared by Sol-Gel Pyrolysis Method", *IEEE Translation Journal on Magnetics in Japan*, vol. 6, no. 8, pp. 648–653, Aug. 1991.

[31] W. Zhong, W. Ding, N. Zhang, J. Hong, Q. Yan, and Y. Du, "Key Step in Synthesis of Ultrafine $BaFe_{12}O_{19}$ by Sol-Gel Technique", *Journal of Magnetism and Magnetic Materials*, vol. 168, no. 1, pp. 196–202, Apr. 1997.

[32] L. Wen-Tsang and W. Jenn-Ming, "The Effect of the Vacuum Extraction and the Fe/Ba Ratio on the Phase Formation of Barium Ferrite Thin Film Synthesized by Sol-Gel Method", *Materials Chemistry and Physics*, vol. 69, pp. 148–153, Mar. 2001.

[33] S. Y. An, S. W. Lee, I.-B. Shim, S.-R. Yun, and C. S. Kim, "Barium Hexaferrite Thin Films Prepared by the Sol-Gel Method", *Journal of Magnetics*, vol. 6, no. 1, pp. 23–26, Mar. 2001.

[34] S. Y. An, S. W. Lee, I.-B. Shim, and C. S. Kim, "Magnetic Properties of Water-Based Sol-Gel Derived
$BaFe_{12}/O_{19}/SiO_2/Si(100)$ Thin Films", *IEEE Transactions on Magnetics*, vol. 37, no. 4, pp. 2585–2588, Jul. 2001.

[35] S. Salemizadeh and S. A. S. Ebrahimi, "Magnetic Properties of Nanocrystalline Barium Hexaferrite Thin Films Prepared by Sol-Gel Method", *IEEE Transactions on Magnetics*, vol. 45, no. 6, pp. 2538–2540, May 2009.

[36] H. Li, J. Huang, Q. Li, and X. Su, "Preparation of Barium Ferrite Films with High Fe/Ba Ratio by Sol-Gel Method", *Journal of Sol-Gel Science and Technology*, vol. 52, no. 3, pp. 309–314, Aug. 2009.

[37] H. Tang, W. Zhang, B. Peng, and W. Zhang, "Magnetically Anisotropic Barium Ferrite Thin Films on Sapphire by Sol-Gel Process", *Thin Solid Films*, vol. 518, no. 12, pp. 3342–3344, Apr. 2010.

[38] S. Meng, Z. Yue, and L. Li, "Effect of Ethylene Glycol on the Orientation and Magnetic Properties of Barium Ferrite Thin Films Derived by Chemical Solution Deposition", *Journal of Magnetism and Magnetic Materials*, vol. 354, pp. 290–294, Mar. 2014.

[39] T. Jensen, V. Krozer, and C. Kjærgaard, "Realisation of Microstrip Junction Circulator using LTCC Technology", *Electronics letters*, vol. 47, no. 2, pp. 111–113, Jan. 2011.

[40] T. Jensen, "LTCC Cofired Circulators for Compact Radar Circuits", PhD thesis, DTU Electrical Engineering, Technical University of Denmark, Lyngby, Denmark, 2010.

[41] R. v. Dijk, G. Bent, M. Ashari, and M. McKay, "Circulator Integrated in Low Temperature Co-fired Ceramics Technology", in *9th European Microwave Integrated Circuits Conference (EuMIC 2014)*, Rome, Italy, Oct. 2014, pp. 600–603.

[42] B. E. Warren, *X-Ray Diffraction*. Dover, New York, NY, USA, 1990, Originally published: Reading, Mass.: Addison-Wesley, Boston, MA, USA, 1969.

[43] W. Wong-Ng, H. McMurdie, B. Paretzkin, C. Hubbard, and A. Dragoo, "JCPDS-ICDD Research Associateship, Cooperative Program with NBS/NIST", *Journal of Research of NIST*, vol. 106(6), pp. 1013–1028, 1988.

[44] C. Tenailleau, M. Allix, J. B. Claridge, M. Hervieu, M. F. Thomas, J. P. Hirst, and M. J. Rosseinsky, "Modular Construction of Oxide Structures-Compositional Control of Transition Metal Coordination Environments", *Journal of the American Chemical Society*, vol. 130, pp. 7570–7583, 2008.

[45] L. Pauling and S. B. Hendricks, "Crystal Structures of Hematite and Corundum Locality: Ceylon", *Journal of the American Chemical Society*, vol. 47, pp. 781–790, 1925.

[46] S.-Y. Chen and I.-W. Chen, "Cracking during Pyrolysis of Oxide Thin Films-Phenomenology, Mechanisms, and Mechanics", *Journal of the American Ceramic Society*, vol. 78, no. 11, pp. 2929–2939, Nov. 1995.

[47] F. K. H. Gellersen, F. M. Klichowski, A. G. C. Rusche, and A. F. Jacob, "Barium Hexaferrite Films Prepared by Means of the Sol-Gel Method for Self-Biased Nonreciprocal Microwave Devices", in *2016 German Microwave Conference (GeMiC)*, Bochum, Germany, Mar. 2016, pp. 221–224.

[48] J. Santos, M. Macedo, F. Cunha, J. Sasaki, and J. Duque, "$BaFe_{12}O_{19}$ Thin Film Grown by an Aqueous Sol-Gel Process", *Microelectronics Journal*, vol. 34, no. 5, pp. 565–567, May 2003.

[49] V. N. Dhage, M. Mane, A. Keche, C. Birajdar, and K. Jadhav, "Structural and Magnetic Behaviour of Aluminium Doped Barium Hexaferrite Nanoparticles Synthesized by Solution Combustion Technique", *Physica B: Condensed Matter*, vol. 406, no. 4, pp. 789–793, Nov. 2011.

[50] S. Diaz, J. Sanchez, B. Watts, F. Leccabue, and R. Panizzieri, "Magnetic Properties of Polycrystalline $BaFe_{12}O_{19}$ Thin Films Prepared from Metallorganic Decomposition on ZrO_2-Coated Si Substrates", *Journal of Magnetism and Magnetic Materials*, vol. 151, no. 1-2, pp. 173–177, Nov. 1995.

[51] C. Wen, "Coplanar Waveguide: A Surface Strip Transmission Line Suitable for Non-reciprocal Gyromagnetic Device Applications", *IEEE Transactions on Microwave Theory and Techniques*, vol. 17, no. 12, pp. 1087–1090, Dec. 1969.

[52] F. K. H. Gellersen, J. Peschel, A. Ochsenfarth, and A. F. Jacob, "Scandium-Doped Barium Hexaferrite Thin-Films for Nonreciprocal Satellite Components", in *2017 Topical Workshop on Internet of Space (TWIOS)*, Jan. 2017, pp. 1–4.

[53] S. H. Mahmood, G. H. Dushaq, I. Bsoul, M. Awawdeh, H. K. Juwhari, B. I. Lahlouh, and M. A. AlDamen, "Magnetic Properties and Hyperfine Interactions in M-Type $BaFe_{12-2x}Mo_xZn_xO_{19}$ Hexaferrites", *Journal of Applied Mathematics and Physics*, pp. 77–87, Apr. 2014.

[54] J. Qiu, Q. Zhang, M. Gu, and H. Shen, "Effect of Aluminum Substitution on Microwave Absorption Properties of Barium Hexaferrite", *Journal of Applied Physics*, vol. 98, no. 10, p. 103 905, Oct. 2005.

[55] A. Yang, Y. Chen, Z. Chen, C. Vittoria, and V. G. Harris, "Magnetic and Atomic Structure Parameters of Sc-Doped Barium Hexagonal Ferrites", *Journal of Applied Physics*, vol. 103, no. 7, pp. 07E511, Mar. 2008.

[56] S. Bierlich, F. Gellersen, A. Jacob, and J. Töpfer, "Low-Temperature Sintering and Magnetic Properties of Sc- and In-Substituted M-Type Hexagonal Barium Ferrites for Microwave Applications", *Materials Research Bulletin*, vol. 86, pp. 19–23, Feb. 2017.

[57] S. Oliver, M. Chen, C. Vittoria, and P. Lubitz, "Properties of Pulsed Laser Deposited Scandium-Doped Barium Hexaferrite Films", *Journal of Applied Physics*, vol. 85, no. 8, pp. 4630–4632, Apr. 1999.

[58] P. Shi, S. Yoon, X. Zuo, I. Kozulin, S. Oliver, and C. Vittoria, "Microwave Properties of Pulsed Laser Deposited Sc-Doped Barium Hexaferrite Films", *Journal of Applied Physics*, vol. 87, no. 9, pp. 4981–4983, Apr. 2000.

[59] V. Harris, Y. Chen, and Z. Chen, "Crystallographic Texture and Magnetic Anisotropy and Their Influence Upon Microwave Devices", *JOM*, vol. 65, no. 7, pp. 883–889, Jun. 2013.

[60] F. K. H. Gellersen, C. Diker, J. Peschel, A. Ochsenfarth, and A. F. Jacob, "Doped Barium Hexaferrite Films for Self-Biased Nonreciprocal Components in the Q-/V-Band", *IEEE Antennas and Wireless Propagation Letters*, vol. 17, no. 11, pp. 1938–1942, Nov. 2018.

[61] D. Vinnik, D. Zherebtsov, L. Mashkovtseva, S. Nemrava, M. Bischoff, N. Perov, A. Semisalova, I. Krivtsov, L. Isaenko, G. Mikhailov, *et al.*, "Growth, Structural and Magnetic Characterization of Al-Substituted Barium Hexaferrite Single Crystals", *Journal of Alloys and Compounds*, vol. 615, pp. 1043–1046, Dec. 2014.

[62] A. Ustinov, A. Tatarenko, G. Srinivasan, and A. Balbashov, "Al Substituted Ba-Hexaferrite Single-Crystal Films for Millimeter-Wave Devices", *Journal of Applied Physics*, vol. 105, no. 2, p. 023 908, Jan. 2009.

[63] D. Vincent, T. Rouiller, C. Simovsky, B. Bayard, and G. Noyel, "A New Broad-Band Method for Magnetic Thin-Film Characterization in the Microwave Range", *IEEE Trans. Microwave Theory Techn.*, vol. 53, no. 4, pp. 1174–1180, Apr. 2005.

[64] S. Yoon, C. Vittoria, and S. Oliver, "Magnetic and Microwave Magnetic Properties of Barium Hexaferrite Permanent Magnet Films having the *c*-Axis in the Film Plane", *Journal of applied physics*, vol. 93, no. 7, pp. 4023–4026, Mar. 2003.

[65] Z. Zhuang, M. Rao, R. M. White, D. E. Laughlin, and M. H. Kryder, "Barium Ferrite Thin Film Media with Perpendicular *c*-Axis Orientation and Small Grain Size", *Journal of Applied Physics*, vol. 87, no. 9, pp. 6370–6372, Apr. 2000.

[66] D. Chen, Z. Chen, G. Wang, Y. Chen, and Y. Li, "Magnetic Properties and Millimeter Wave Loss of Highly Oriented Scandium Substituted Barium Hexaferrite Thin Films for Millimeter Wave Applications", *Journal of Materials Science: Materials in Electronics*, vol. 28, no. 9, pp. 6737–6740, May 2017.

[67] Y. Chen and M. H. Kryder, "Temperature Dependent Magnetic Properties of Barium-Ferrite Thin-Film Recording Media", *IEEE Transactions on Magnetics*, vol. 34, no. 3, pp. 729–742, May 1998.

[68] T. Rabe, M. Gemeinert, and W. A. Schiller, "Development of Advanced Low Temperature Co-Fired Ceramics (LTCC)", in *Key Engineering Materials*, Trans. Tech. Publ., vol. 264, May 2004, pp. 1181–1184.

[69] S. Dai, "Densification and Crystallization in Crystallizable Low Temperature Co-Fired Ceramics", *Journal of Materials Science*, vol. 47, no. 11, pp. 4579–4584, Jun. 2012.

[70] M. T. Sebastian, R. Ubic, and H. Jantunen, *Microwave Materials and Applications*, 1st E., ser. Wiley series in materials for electronic & optoelectronic applications. John Wiley & Sons, Hoboken, NJ, USA, Mar. 2, 2017.

[71] A. Mohanram, G. L. Messing, and D. J. Green, "Densification and Sintering Viscosity of Low-Temperature Co-Fired Ceramics", *Journal of the American Ceramic Society*, vol. 88, no. 10, pp. 2681–2689, Jun. 2005.

[72] J. Zhou, "Towards Rational Design of Low-Temperature Co-Fired Ceramic (LTCC) Materials", *Journal of Advanced Ceramics*, vol. 1, no. 2, pp. 89–99, Jun. 2012.

[73] M. Gemeinert, "Über LTCC-Werkstoffe aus dem Stoffsystem CaO-La_2O_3-Al_2O_3-B_2O_3", PhD thesis, BAM Bundesanstalt für Materialforschung und -prüfung, Berlin, Germany, Feb. 2009.

[74] H. Yu, J. Liu, W. Zhang, and S. Zhang, "Ultra-Low Sintering Temperature Ceramics for LTCC Applications: a Review", *Journal of Materials Science: Materials in Electronics*, vol. 26, no. 12, pp. 9414–9423, Dec. 2015.

[75] W. K. Jones, Y. Liu, B. Larsen, P. Wang, and M. Zampino, "Chemical, Structural and Mechanical Properties of the LTCC Tapes", in *Proceedings-SPIE the international society for optical engineering*, International Society for Optical Engineering; 1999, vol. 23 No. 4, pp. 669–674.

[76] A. Talai, F. Steinhäußer, A. Bittner, U. Schmid, R. Weigel, and A. Koelpin, "A Permittivity Characterization Method by Detuned Ring-Resonators for Bulk Materials up to 110 GHz", in *44th European Microwave Conference (EuMC)*, IEEE, Rome, Italy, Oct. 2014, pp. 124–127.

[77] A. Allanore, G. Lambotte, and J. Lee, *Materials Processing Fundamentals 2017*, ser. The Minerals, Metals & Materials Series. Springer, Berlin, Germany, Mar. 2017.

[78] P. Scherrer, "Bestimmung der Größe und der inneren Struktur von Kolloidteilchen mittels Röntgenstrahlen", *Nachrichten von der Gesellschaft der Wissenschaften zu Göttingen*, vol. 2, pp. 98–112, Jul. 1918.

[79] V. Laur, G. Vérissimo, P. Quéffélec, H. Alaaeddine, J.-C. Reihs, E. Laroche, G. Martin, R. Lebourgeois, J.-P. Ganne, *et al.*, "Modeling and Characterization of Self-Biased Circulators in the mm-Wave Range", in *IEEE MTT-S International Microwave Symposium (IMS)*, IEEE, Phoenix, AZ, USA, May 2015.

[80] V. Laur, G. Vérissimo, P. Quéffélec, L. A. Farhat, H. Alaaeddine, E. Laroche, G. Martin, R. Lebourgeois, and J. Ganne, "Self-Biased Y-Junction Circulators Using Lanthanum- and Cobalt-Substituted Strontium Hexaferrites", *IEEE Transactions on Microwave Theory and Techniques*, vol. 63, no. 12, pp. 4376–4381, Dec. 2015.

[81] S. Oliver, P. Shi, N. McGruer, C. Vittoria, W. Hu, H. How, S. McKnight, and P. Zavracky, "Integrated Self-Biased Hexaferrite Microstrip Circulators for Millimeter-Wavelength Applications", *IEEE Transactions on Microwave Theory and Techniques*, vol. 49, no. 2, pp. 385–387, Feb. 2001.

[82] B. K. O'Neil and J. L. Young, "Experimental Investigation of a Self-Biased Microstrip Circulator", *IEEE Transactions on Microwave Theory and Techniques*, vol. 57, no. 7, pp. 1669–1674, May 2009.

[83] B. Peng, H. Xu, H. Li, W. Zhang, Y. Wang, and W. Zhang, "Self-Biased Microstrip Junction Circulator Based on Barium Ferrite Thin Films for Monolithic Microwave Integrated Circuits", *IEEE Transactions on Magnetics*, vol. 47, no. 6, pp. 1674–1677, Jun. 2011.

[84] J. Wang, A. Yang, Y. Chen, Z. Chen, A. Geiler, S. M. Gillette, V. G. Harris, and C. Vittoria, "Self Biased Y-Junction Circulator at K_U Band", *IEEE Microwave and Wireless Components Letters*, vol. 21, no. 6, pp. 292–294, May 2011.

[85] X. Zuo, H. How, S. Somu, and C. Vittoria, "Self-Biased Circulator/Isolator at Millimeter Wavelengths Using Magnetically Oriented Polycrystalline Strontium M-Type Hexaferrite", *IEEE Transactions on Magnetics*, vol. 39, no. 5, pp. 3160–3162, Sep. 2003.

[86] F. K. H. Gellersen and A. F. Jacob, "A Sol-Gel Approach for Self-Biased Barium Hexaferrite Thin-Film Circulators", in *46th European Microwave Conference (EuMC)*, London, United Kingdom, Oct. 2016, pp. 361–364.

[87] H. Bosma, "On stripline Y-circulation at UHF", *IEEE Transactions on Microwave Theory and Techniques*, vol. 12, no. 1, pp. 61–72, Jan. 1964.

[88] S. Yang, L. Roy, J. R. Bray, and D. Vincent, "Modeling and Simulation of a Partially-Magnetized Ferrite LTCC Circulator", in *IEEE MTT-S International Conference on Numerical Electromagnetic and Multiphysics Modeling and Optimization (NEMO)*, Ottawa, ON, Canada, Aug. 2015.

[89] G. Strauss, P. Ehret, and W. Menzel, "On-Wafer Measurement of Microstrip-Based MIMICs Without Via Holes", in *IEEE MTT-S International Microwave Symposium Digest*, vol. 3, San Francisco, CA, USA, Jun. 1996.

Lebenslauf

Name	Gellersen
Vorname	Frauke Kathinka Helene
Geburtsdatum	22.09.1988
Geburtsort, -land	Winsen (Luhe), Deutschland

08.1995 - 07.1999	Grundschule Salzhausen-Eyendorf in Salzhausen
08.1999 - 07.2001	Orientierungsstufe Salzhausen
08.2001 - 07.2005	Oberschule Salzhausen
08.2005 - 07.2008	Technisches Gymnasium der Georg-Sonnin-Schule in Lüneburg
10.2008 - 07.2011	Studium an der Technischen Universität Hamburg-Harburg in Hamburg Abschluss: Bachelor of Science
10.2011 - 09.2013	Studium an der Technischen Universität Hamburg-Harburg in Hamburg Abschluss: Master of Science
09.2013 - 07.2017	Wissenschaftliche Mitarbeiterin an der Technischen Universität Hamburg-Harburg in Hamburg
08.2017 - heute	Hardware-Entwicklerin bei Panasonic Industrial Devices Europe in Lüneburg